Building Pathology and Rehabilitation

Volume 18

This book series addresses the areas of building pathologies and rehabilitation of the constructed heritage, strategies, diagnostic and design methodologies, the appropriately of existing regulations for rehabilitation, energy efficiency, adaptive rehabilitation, rehabilitation technologies and analysis of case studies. The topics of Building Pathology and Rehabilitation include but are not limited to - hygrothermal behaviour - structural pathologies (e.g. stone, wood, mortar, concrete, etc...) - diagnostic techniques - costs of pathology - responsibilities, guarantees and insurance - analysis of case studies - construction code - rehabilitation technologies - architecture and rehabilitation project - materials and their suitability - building performance simulation and energy efficiency - durability and service life.

More information about this series at http://www.springer.com/series/10019

João M. P. Q. Delgado
Editor

Building Pathologies and Acoustic Performance

Springer

Editor
João M. P. Q. Delgado
CONSTRUCT-LFC
Department of Civil Engineering
Faculty of Engineering
University of Porto
Porto, Portugal

ISSN 2194-9832 ISSN 2194-9840 (electronic)
Building Pathology and Rehabilitation
ISBN 978-3-030-71232-7 ISBN 978-3-030-71233-4 (eBook)
https://doi.org/10.1007/978-3-030-71233-4

This Springer imprint is published by the registered company Springer Nature Switzerland AG
The registered company address is: Gewerbestrasse 11, 6330 Cham, Switzerland

Preface

During the last years, a great intensity of new buildings construction was occurred, representing these buildings an enormous patrimonial valour, which require a permanent investment in management, maintenance, repair and rehabilitation.

Rehabilitation investment in the built environment, beyond contributing for the preservation of the existing patrimony, contributes for the implementation of construction sustainable politics, through the reduction of the consumption of natural resources and energy, of the reduction of residues proceeding from the construction activity and for one better management of ground occupation. In opposition to this, has been registered a quality and durability decrease of the public and private built environment through its precocious degradation. The aim of this book is to identify some anomalies that affect the acoustic performance and the concrete structures, namely, anomalies associated to salt attack and corrosion that affects the durability and service life of the buildings, in order to be eliminated in future projects.

The main purpose of this book, *Building Pathologies and Acoustic Performance*, is to provide a collection of recent research works related to salt attack and corrosion in concrete structures, durability and service life in concrete, non-destructive techniques applied to reinforced concrete structures and acoustic performance of buildings.

The book is divided into five chapters that intend to be a resume of the current state of knowledge for benefit of professional colleagues, scientists, students, practitioners, lecturers and other interested parties to network. At the same time, these topics will be going to the encounter of a variety of scientific and engineering disciplines, such as civil, mechanical and materials engineering.

Porto, Portugal João M. P. Q. Delgado

Contents

Making Safer Cities from Damaged Foot-Over-Bridges

Prafulla Parlewar

Abstract The urban foot-over-bridges are part of everyday life of citizens which provide access to transportation. In Mumbai, India on September 26, 2017 a major stampede happened on a railway foot over bridge. In many cities, the old foot-over-brigades are heavily damaged which may cause a fatal accident. How can we analyse a foot-over-bridge to understand level of safety? Which NDT techniques can provide us accurate analysis? How can we make a planning for a common policy to make our cities safer? It is indeed important to investigate the damages of foot-over-bridges to make our cities safer to avoid stampede. Some of the important aspects of the investigation of these foot-over-bridges are the strength of the concrete, damages of the steel element and severity in damages in the steel connections which causes the failure. Further in this research, a case study is illustrated of railway foot-over-bridge deteriorated with passage of time. The investigation found that the foot-over-bridge was completely damaged and need an urgent demolition for the safety of the citizens.

Keywords Safer cities · Foot-over-bridges · NDT testing

1 Introduction

Cities across globe are gradually becoming pedestrian and bicycle-friendly. Cities in Japan are planned on walk-friendly concepts, and a city like Copenhagen has become an ideal model for pedestrian and bicycle city. Foot-over-bridge is a very important component of pedestrian-friendly cities. It connects two points of destinations for easy access, mainly, the railway foot-over-bridges provide access to cities for fast commuting from one place to another. In South Asia, particularly in Indian cities, railway foot-over-bridges were constructed before 1945 during the British period. These old bridges are now deteriorated due to the weather and less carry capacity. From, the year 2018 to 2020, three major stampedes happen in the city of Mumbai. This accident affected many citizens. After these stampedes, the local government

P. Parlewar (✉)
City Development Corporation (P) Ltd., Mumbai, India

J. M. P. Q. Delgado (ed.), *Building Pathologies and Acoustic Performance*,
Building Pathology and Rehabilitation 18, https://doi.org/10.1007/978-3-030-71233-4_1

and railways started auditing the bridges to find the problems. So, How can we analyze a foot-over-bridge to understand the level of safety? Which NDT techniques can provide us accurate analysis? How can we make a planning for a common policy to make our cities safer? It is indeed important to investigate the damages of foot-over-bridges to make our cities safer to avoid the stampede. Some of the important aspects of the investigation of these foot-over-bridges are the strength of the concrete, damages of the steel element, and severity in damages in the steel connections which causes the failure. The foot over bridges are small but important to avoid fatal accidents and traffic congestion, efficiently convey the traffic, ultimately decrease the destination time (Chandrikka et al. 2019).

Several bridges in India are constructed few decades ago, are not serviceable and requires immediate attention due to damage and deterioration (Shukla and Maiti 2019). Many cities in India are facing these problem. Urban Planner and engineers are identifying the solution for making safer cities. The condition assessment of an existing bridge requires a very complex and comprehensive approach deeply involving both the use of standard procedures, like in situ and laboratory tests (by which the characterization of material properties and of the principal effects of deterioration phenomena can be obtained) and less conventional structural analyses like structural monitoring (Modena et al. 2012).

Structural stability analysis of the bridge can be conducted in three parts: (a) geometric studies, (b) structural testing and (c) vibration analysis. A typical case study of railway foot-over-bridge of 250 m long and 1.8 m wide of simply supported type indicate in geometric studies the damages in concrete and steel member. The steel connections of the bridge were found to be damaged critically. Various types of tests are recommended for the foot-over-bridge. These tests include the rebound hammer test, ultrasonic pulse velocity test, carbonation depth measurement test, core cutting, and compression test, dye penetration test, and safe bearing capacity test. Also, sometimes load tests clarify the doubts about the shear strength. But it can also be used for the analysis of the problems of the flexural capacity of concrete (Parlewar 2021b). The case study foot-over-bride was found to be the critical condition in the superstructure. So, it was recommended to demolish the bridge and stop any further use by pedestrians.

2 Geometric Studies

The existing railway foot-over-bridge is 250 m long bridge located on railway lines. The bridge is designed with R.C.C. foundation and steel superstructure. A R.C.C. slab is supported on the steel structure for pedestrian movement. The width of the pathway is 1.8 m wide. The R.C.C. is 200 mm thick. The steel structure is mainly comprised of columns in box section ($2 \times$ ISMC 100) with cross double MS angle ($2 \times$ ISA $75 \times 75 \times 6$) bracing. The bottom of the slab is supported with ISMB 450. Column supports are equally placed at a distance of 4.5 m (Figs. 1 and 2). The geometric studies indicated that the design of foot-over-bridge is simply supported steel

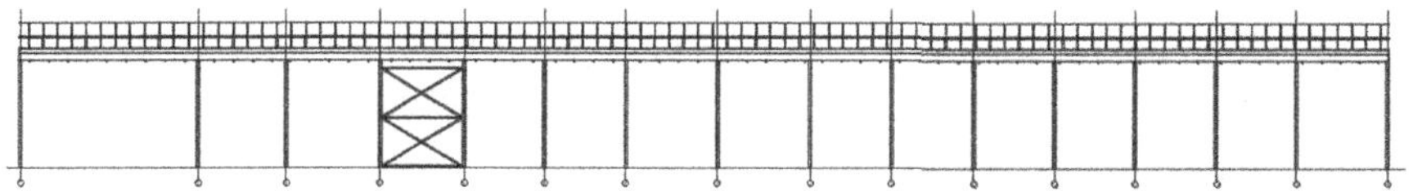

Fig. 1 Elevation of foot-over-bridge

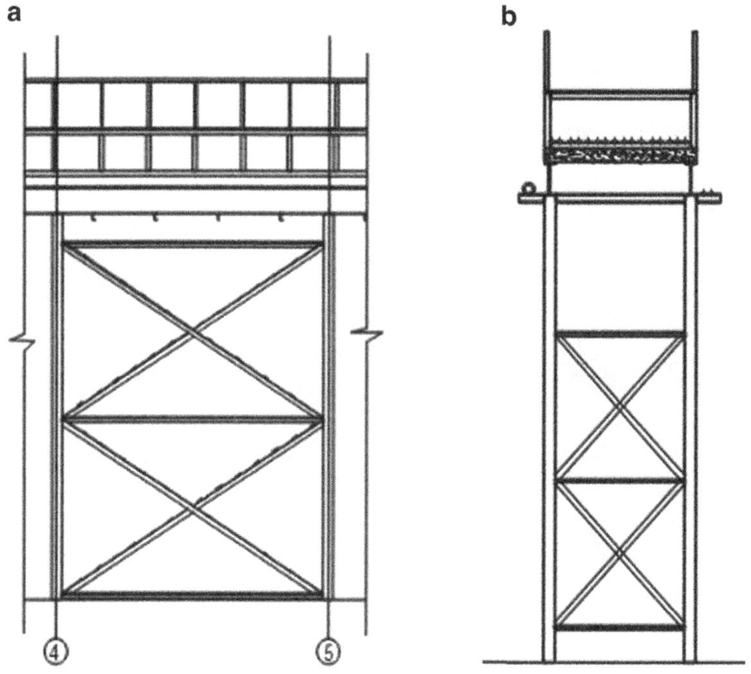

Fig. 2 **a** Election of column and **b** section of foot-over-bridge

structure. The foot-over-bridge is subjected to load of 1500 persons per hour traffic. Generally, the peak time is in morning 9.00 am to 11.00 am and evening 5.00 pm to 7.00 pm. The bridge is subjected to the vibration by the railways passing underneath. As the design of the bridge is very old, care is not taken to provide safety from the traffic of railway passing underneath.

3 Visual Analysis

The visual analysis had indicated that the steel frame is severely damaged over the bridge (Figs. 3, 4, 5, 6, 7 and 8). The steel connection is found to be damaged due to weather conditions. Also, severe damages like corrosion are found in the steel structure of the foot-over-bridge. Along with the steel structure, the concrete was also found not in good condition.

Fig. 3 Views of connections

Fig. 4 Damages in steel connections

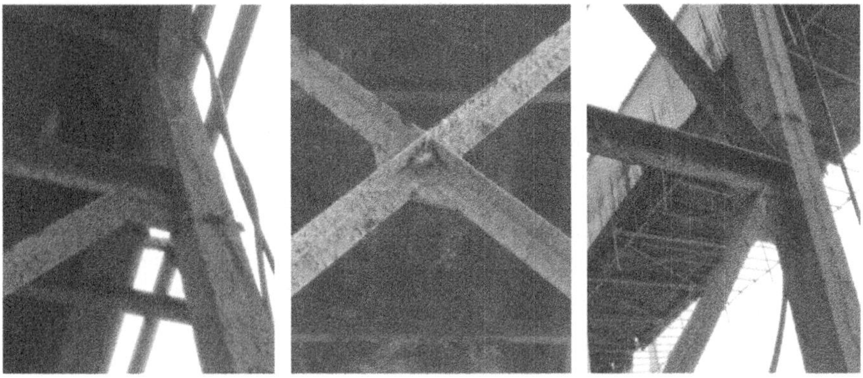

Fig. 5 Damages in steel connections

Fig. 6 Damages in steel connections

Fig. 7 Damages in steel connections

Fig. 8 Damages in steel connections

4 Structural Testing

4.1 Rebound Hammer Test

The rebound hammer test is conducted as per the IS 13311 Part 2 and BS 1881 Part 202. The rebound hammer test is conducted on the principle of rebound of the plunger held against the concrete surface. To undertake this test the plunger is pressed against the surface of the concrete. The spring inside the equipment rebounds to the extent of the hardness of the concrete. The reading is then correlated with the graph on the instrument as the compressive strength of the concrete. A variation of 25% is found in the strength of the specimen tested by RHT. This is because of factors like surface and internal moisture, carbonation of concrete, age of concrete, and type of aggregates (Parlewar 2021a).

The purpose of the rebound hammer test is to assess the compressive strength of the concrete. The test also helps to assess the uniformity of the concrete. Furthermore, the concrete compressive strength can be correlated with the standards compressive strength of the concrete.

The procedure for the rebound hammer test include first cleaning of the surface. In the case of loose mortar, it is necessary to clean and remove the mortar before taking the test. Normally, the test can be conducted by holding the plunger at 45° or 90° as indicated on the graph. Generally, the test is conducted by taking nine points on the concrete surface. These, nine points are marked in a rectangular grid with three on top, three in the middle, and three on the bottom average of these nine points is considered the rebound reading for estimating the compressive strength of concrete.

Rebound Hammer Test results shows compressive strength between 34–54 Mpa of foot-over-bridge (Fig. 9 and Table 1). As per the IRC code, the compressive strength of concrete required is above 35 Mpa. The concrete strength is found to be below 35 Mpa. This infers that the foundation is not in good condition.

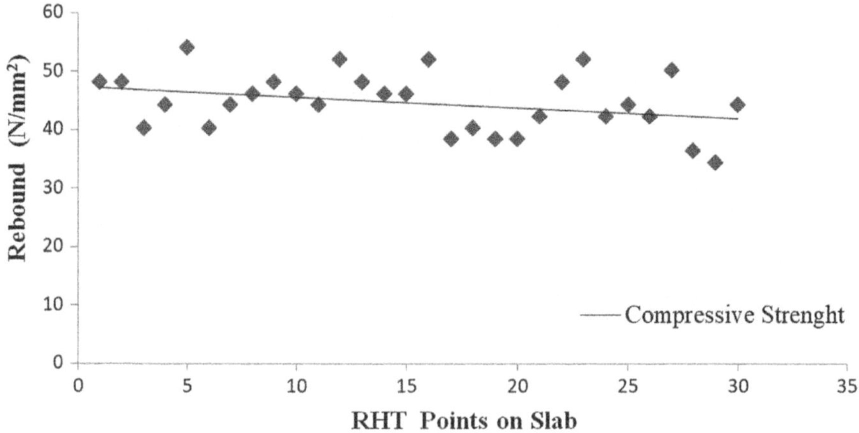

Fig. 9 Results of Rebound Hammer test

Table 1 Results of Rebound Hammer test

S/N	Member	Average compressive strength (Mpa)	Remark
1	PS1	34.32	Poor
2	PS2	34.32	Poor
3	PS3	40.21	Average
4	PS4	44.13	Average
5	PS5	53.94	Good
6	PS6	40.21	Average
7	PS7	44.13	Average
8	PS8	46.09	Good
9	PS9	48.05	Good
10	PS10	46.09	Good
11	PS11	44.13	Average
12	PS12	51.98	Good
13	PS13	48.05	Good
14	PS14	46.09	Good
15	PS15	46.09	Good
16	PS16	51.98	Good
17	PS17	38.25	Average
18	PS18	38.25	Average
19	PS19	42.17	Poor
20	PS20	48.05	Poor
21	PS21	51.98	Average
22	PS22	42.17	Good
23	PS23	44.13	Good
24	PS24	42.17	Average
25	PS25	44.13	Average
26	PS26	42.17	Average
27	PS27	50.01	Good
28	PS28	36.28	Poor
29	PS29	34.32	Poor
30	PS30	44.13	Average
31	PS31	44.13	Average

4.2 Ultrasonic Pulse Velocity Test

The Ultrasonic Pulse Velocity Test (UPV) is conducted as per the IS 516 (Part/Sec 1) IS 2018, IS 13311 Part 1, BS 1881 Part 203. The ultrasonic pulse is generated by an electro-acoustical transducer. The pulse is inserted into the concrete by holding the two ends of the UPV testing instrument. The pulse undergoes multiples reflection at the end of the various materials inside the concrete. Stress waves are generated in three types: (a) longitudinal (compression), shear (transverse), and surface (Rayleigh) waves.

The UPV test identifies the homogeneity of the concrete. It identifies the presence of cracks, voids, and any other defects in the concrete. These investigations are important to under the changes in the concrete structure with the passage of time. It is expected that concrete changes its properties due to the weather, and vibration. So, there is the possibility of porosity and damages to the internal structure of concrete. Importantly, to maintain the safety of the foot-over-bridges, minimum strength criteria have to be maintained. So, UPV provides information about the strength of the concrete in comparison with the standards code for concrete structures. UPV test also makes it easy to compare various concrete elements for its strengths. This comparison investigates the present situation of the concrete members and requirements for the retrofitting. Moreover, the UPV test explains the dynamic elastic modulus of the concrete. The procedure of the UPV test includes first cleaning the concrete surface. The cleaning is conducted by scrapping and bruising the surface before using the UPV equipment. Then, grease is applied on the concrete surface at the locations of test points. The instrument probe is pressed firm to remove any air gaps in the surface of concrete. The distance between two probes (D) is noted and time (T) checked on the instrument. The velocity is calculated with formula V equal to D/T. This test is continued on the location identified for the test. A chart is prepared and valued are checked as per the Table 4. The UPV test results are included by various factors like

Table 2 Study of UPV test on R.C.C. floor of bridge

S/N	Member	Member ID	Level	Element type	UPV (Kms)
1	FT. OB/TOP	S1	FT.OB	Slab	3.21
2	FT. OB/TOP	S2	FT.OB	Slab	3.13
3	FT. OB/Bottom	S3	FT.OB	Slab	2.33
4	FT. OB/Bottom	S4	FT.OB	Slab	2.45
5	FT. OB/Bottom	S5	FT.OB	Slab	3.12
6	FT. OB/Bottom	S6	FT.OB	Slab	1.83
7	FT. OB/Bottom	S7	FT.OB	Slab	2.11
8	Foundation	FC1	F.D.	Column	3.18
9	Foundation	FC2	F.D.	Column	3.14
10	Foundation	FC3	F.D.	Column	3.22

Table 3 Results of UPV test on R.C.C. floor of bridge

S/N	Member	Member ID	Concrete quality	Depth (mm)
1	FT. OB/TOP	S1	Doubtful	35
2	FT. OB/TOP	S2	Doubtful	30
3	FT. OB/Bottom	S3	Poor	30
4	FT. OB/Bottom	S4	Poor	25
5	FT. OB/Bottom	S5	Doubtful	25
6	FT. OB/Bottom	S6	Poor	25
7	FT. OB/Bottom	S7	Poor	30
8	Foundation	FC1	Doubtful	50
9	Foundation	FC2	Doubtful	40
10	Foundation	FC3	Doubtful	55

Table 4 UPV result criteria and summary

S/N	Criteria	Concrete quality	Reading
1	Above 4.40 km/s	Excellent	0
2	3.75 to 4.40 km/s	Good	0
3	3.00 to 9.75 km/s	Doubtful	0
4	Below 3.0 km/s	Poor	4

smoothness of surface, size of concrete, shape of concrete, moisture, temperature and reinforcement inside the concrete.

The Ultrasonic Pulse Velocity (UPV) test indicated that the concrete quality is doubtful and poor at six tested locations on the R.C.C. Slab. Also, the concrete quality is doubtful on the foundations. So, it can be inferred that the concrete slab and foundation are in critical condition (Tables 2 and 3).

4.3 *Carbonation Depth Measurement Test*

Carbonation of concrete happens when Carbon di Oxide in concrete reacts with Calcium Hydro-oxide to form Calcium Carbonate. To undertake this test the Phenolphthalein solution is applied to the concrete sample. This test indicates the corrosion in the reinforcement. Concrete becomes weak in strength if the corrosion is high.

The pH values for the slab are 9 and 10. The pH values for the foundations are 9, 9, and 11. Even though pH values of 9–11 appear to be close to 8.5, it should be noted that a pH value of 10 could be considered carbonated concrete, while a pH of 12 could be considered non-carbonated concrete (Tables 5 and 6).

So, it can be inferred that the result of pH test shows carbonation in concrete. Moreover, it can be inferred that the R.C.C. members are not in good condition.

Table 5 Results of carbonation test on R.C.C. members

S/N	Member	Member ID	Level	Element type	pH	CL (kgm^3)	SO^3(%)
1	Slab	S1	FT.OB	SLAB	10	0.389	<0.5
2	Slab	S2	FT.OB	SLAB	9	0.41	<0.5
3	Foundation	C1	G.F.	F.D.	11	0.45	<0.5
4	Foundation	C2	G.F.	F.D.	9	0.466	<0.5
5	Foundation	C3	G.F.	F.D.	9	0.456	<0.5

Table 6 Permissible limits as per IS 456:2000

Item	Permissible limit
pH	Not <8.5
CL (kg/m^3)	For PCC, Max. 3.0 kg of Chloride per M^3 of concrete
CL (kg/m^3)	For RCC, Max. 0.6 kg of Chloride per M^3 of concrete
SO^3 (%)	<4% by mass of cement in the concrete mix

4.4 Core Cutting and Compression Test

The core cutting compression test is conducted as per the IS 456, IS 516, IS 119, BS 1881 part 120. Generally, the core is cut with a machine with a diameter of 755, 100, or 150 mm. Before cutting the core sample the cover meter instrument is used to locate the rainforest to avoid its cutting by the core machine. The core is cut excluding the reinforcement. Thus, it reduces the damages and strength to the slab. After cutting, the core samples are cleaned in lab and kept in water for 48 h. The compression testing methods are used for determining the equivalent cube strength of the concrete. The core cutting and compression test give an assessment of the strength of concrete. The compression of the concrete can also be studied by understanding the color and compression of the concrete members. This is a useful test for identifying the possible cause of the failure in a foot-over-bridges. As the core sample gives the clear study of the failure, these methods provided accurate studies on the failure of concrete members in a foot-over-bridges.

The procedure for the core cutting and compression test is as follows: (1) Reinforcement Mapping-The reinforcement is mapped by the use of special instruments called core meter. The significance of this mapping is to reduce the damages to the concrete member. (2) Core cutting-A diamond cutter is used to cut the core. After identifying the location, the machine is installed and watered at regular intervals to cut the core sample. The cut sample is carefully kept. So, no damage occurs in the sample till it reaches the laboratory. (3) Dressing-At the laboratory, the edges of the core cleaned. Then they are made smooth with help of a grinder or cutter. Later, an epoxy mortar is applied on both sides i.e. top and bottom. The purpose of this epoxy mortar is to make an axis approximately 90°. Also, to ensure that the plainness is within 0.05 mm. (4) Immersion in water-The core samples are immersed in water for 48 h. The core sample weight is taken before and after immersion in water. (5) Compression resting-The core is kept under the compression testing machine to take the reading at the breaking point. Based on this, the equivalent cube strength of the concrete is identified.

Normally, most of the railway foot-over-bridges use concrete as s flooring materials. The concrete floors are damaged by the dynamic pressure of pedestrians and the vibration of the vehicle and rails. Furthermore, weather conditions also deteriorated the concrete. The core cutting and compression test provide an effective method to analyze the foot-over-bridge to avoid a fatal accident.

The reduction in the corrected core strength over the equivalent core strength is 74.48%. The following are the criteria for accepting the strength of the core of the concrete members: (a) the average corrected core strength shall not be <85% of the equivalent cube strength and (b) the strength of the individual member shall not be less than the 75% of equivalent cube strength (Parlewar 2021c). As the average value is below 85%, it can be inferred that the foundation is not in good condition.

4.5 Ultrasonic Thickness Gauge Test

The ultrasonic thickness gauge test is a good indication of the deterioration of the structure. In many old foot-over-bridges with the passage of time, thickness are found reduced of steel members. As the thickness of the steel members is reduced, the member strength is also reduced. It was found that in many locations the members are in a critical situations.

The Ultrasonic Thickness Gauge Test shows that most of the steel members are reduced by more than 20% (Tables 7 and 8). The allowable reduction in thickness may be around 5%. In industrial buildings, the reduction of thickness is 0.075 mm per year. The corrosion allowance of 1.0–1.5 may have been applicable to the steel members. However, it is clear that as the thickness of the steel member is reduced, it directly affects the load-carrying capacity of the steel member.

So, it can be inferred that the member above 20% is in critical condition. Hence, the steel structure is not capable of load carrying capacity. The average reduction in thickness is 18.1% (mm) on the tested members. So, it can be inferred that many of the steel members have problems of corrosion.

4.6 Dye Penetration Test

The dye penetration test can be an important test in the case of analysis of the failure of foot-over-bridges. This test gives knowledge of the weld between the two steel members. This test is mainly based on capillary action. A liquid penetrate is applied on the welded joins and after some time the surface is inspected under the ultraviolet

Table 7 Study of Ultrasonic Thickness Gauge (UTG) on steel members

S/N	Member	Member	Level	Location	Standards (mm)	UTG (mm)
1	IMC175	G1	GF-Column	Center	6.00	5.40
2	IMC175	G2	GF-Column	Center	6.00	4.12
3	IMC175	G3	GF-Column	Center	6.00	6.23
4	IMC175	G4	GF-Column	Center	6.00	5.20
5	IMC175	G5	GF-Column	Center	6.00	4.79
6	IMC175	G6	GF-Column	Center	6.00	6.35
7	IMC175	G7	GF-Column	Center	6.00	5.32
8	IMC175	G8	GF-Column	Center	6.00	4.44
9	IMC175	G9	GF-Column	Center	6.00	5.90
10	IMC175	G10	GF-Column	Center	6.00	5.12
11	IMC175	G11	GF-Column	Center	6.00	5.09
12	IMC175	G12	GF-Column	Center	6.00	6.25
13	IMC175	G13	GF-Column	Center	6.00	4.80
14	IMC175	G14	GF-Column	Center	6.00	4.89
15	IMC175	G15	GF-Column	Center	6.00	5.62
16	IMC175	G16	GF-Column	Center	6.00	6.80
17	IMC175	G17	GF-Column	Center	6.00	4.72
18	IMC175	G18	GF-Column	Center	6.00	5.23
19	IMC175	G19	GF-Column	Center	6.00	6.25
20	IMC175	G20	GF-Column	Center	6.00	5.14

light to underneath the porosity of the weld. In this case study, high porosity was found in the welded joints. Thus, it was concluded that the joints in the bridge are not in good condition (Table 9 and Figs. 10 and 11). Generally, it is not advisable to retrofit these joints. Mainly, because re welding is not possible for these joints.

4.7 Safe Bearing Capacity Test

The safe bearing capacity of the soil is the ability to support the loads imposed by the foot-over-bridges. The test is conducted by excavating a pit in the required soil. Generally, the excavation is done to the depth of the existing foundation. A pressure is dropped on the square pit and then the impression is documented and studied. The safe bearing capacity test result was 13.25175T/m^2. It was found that the solid condition is within the range of the permissible limit of the safe bearing capacity of the foundation.

Table 8 Results of Ultrasonic Thickness Gauge (UTG) on steel members

S/N	Member	Member	Thk. loss (mm)	Thk. loss (%)	Remark
1	IMC175	G1	0.60	10.00	High
2	IMC175	G2	1.88	31.33	Critical
3	IMC175	G3	1.77	22.13	Critical
4	IMC175	G4	0.80	13.33	High
5	IMC175	G5	1.21	20.17	Critical
6	IMC175	G6	1.65	20.63	Critical
7	IMC175	G7	0.68	11.33	High
8	IMC175	G8	1.56	26.00	Critical
9	IMC175	G9	2.10	26.25	Critical
10	IMC175	G10	0.88	14.67	High
11	IMC175	G11	0.91	15.17	High
12	IMC175	G12	1.75	21.88	Critical
13	IMC175	G13	1.20	20.00	Critical
14	IMC175	G14	1.11	18.50	High
15	IMC175	G15	0.38	6.33	Medium
16	IMC175	G16	1.20	15.00	High
17	IMC175	G17	1.28	21.33	Critical
18	IMC175	G18	0.77	12.83	High
19	IMC175	G19	1.75	21.88	Critical
20	IMC175	G20	0.86	14.33	High

Table 9 Result of dye penetration test

S/N	Location	Size (mm)	Type	Remark
1	GF and FL	100	Angle and ISMC	Lack of fusion and cap undercut observed in weld
2	GF	80	Angle and ISMC	Mechanical in weld
3	GF	90	Angle and ISMC	Weld missing, lack of fusion and porosity observed in weld
4	GF	100	ISMC	Lack of sidewall fusion and blow hole seen in weld
5	GF	65	Angle and ISMC	Excess penetration and porocity in weld
6	GF	100	Angle and ISMC	Poor start/stop, lack of root penetration and crater pipe

Fig. 10 Views of dye penetration test

Fig. 11 Views of dye penetration test

5 Vibration Analysis

The vibration analysis was conducted on the foot-over-bridge in STAAD software (Fig. 12). The vibration analysis method included the pedestrian moment on the bridge. Generally, the loading scenario of the design includes pedestrians walking in loads and running loads. The dynamic loads on the pedestrian deck generate the frequencies which affect the concrete and steel structure. The frequencies of the structure depend on the geometry and mechanical properties (Oliveira 2017). The dynamic properties of the structure depend on its mass, its stiffness, and structural damping (Luca et al. 2018). During the loading, these frequencies are considered. Vibration design of footbridges is carried out in order to limit the vertical acceleration

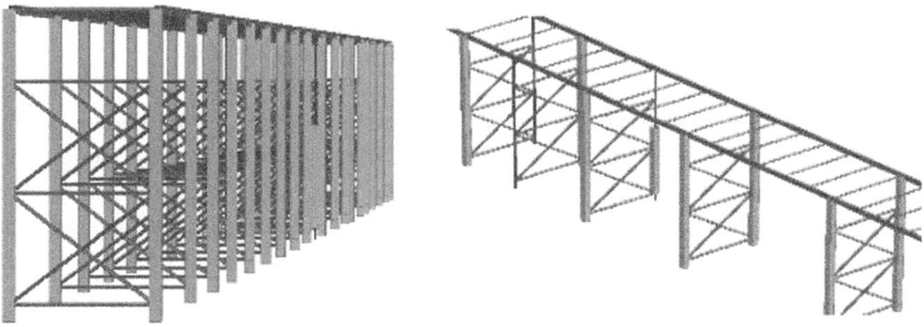

Fig. 12 Views of bridge

of the structure to an acceptable level for the user. The design process is divided into two parts: (i) compute the vertical acceleration, and (ii) ensure it is within acceptable limits, usually referred to as the comfort criteria (Legeron et al. 2000). In the STAAD analysis, many of the members were found to be in the failure state.

The loads acting on the structure is distributed to all the structural elements. The live load and dead load acting on the main truss gets distributed from the gangway to the column. This load is then transferred to the footing below (Rajesh 2017). The wind load acting on foot over the bridge will be comparatively less in magnitude. Because it is an open structure with more openings, but the failure of the towers is mainly due to high-intensity winds and earthquakes. So, a high factor of safety should be given to wind loads and seismic loads (Prashanth and Gokulnath 2018). Some authors have also looked into vandalism and its effect on the vibration of bridges. One of the findings is that the stresses associated with vandal loads remain insignificant when compared with those associated with static loads, as frequently stated in the literature (Caetano et al. 2010). Moreover, not only natural frequencies but also damping properties and pedestrian loading altogether determine the dynamic response (Maca 2017). It may be said that vandal load does not occur often in practice but it certainly deserves greater attention (Zivanovic et al. 2005).

6 Making Safer Cities

The urban pattern in cities includes the movement of peoples from residential to commercial places. So, transportation links these to landuse. To reach work places, peoples commute in high number in the morning and evening on foot-over-bridges. The volume of the commuters are dependent upon the employment opportune in a particular commercial area. In the case of cities line Mumbai, in the gradual process of urbanization, the volume of commuters has increased rapidly and now most of these over bridges are crowded in the morning and evening. So, the planning policies have failed to control the over growth of commercial landuse in the particular commercial area. Hence, this overgrowth is responsible for the overcrowding of the foot-over-bridges. Thus, this bridges leads to deterioration with heavy traffic.

In Mumbai, many of the foot-over-bridges are constructed before 1945. These bridges were constructed by the British without the knowledge of overcrowding populations. The construction technologies used in these bridges were mainly combination and R.C.C. and Steel frames. The bridges must satisfy additional safety and comfort conditions (Salamak and Fross 2016). Some of the bridges are designed as box types and many others are simply supported types. The steel frames are joints with weld which have deteriorated with the passage of time.

Particularly in India, development and management of urban issues are taking by multiples local authorities. The foot-over-bridges is constructed by the railways and local government. This creates an overlay of responsibilities in the management and maintenance of the foot-over-bridges. Many of the commuter's bridges are built by railway authorities. Some of the important policies which cities need to incorporate are (1) planning of the urban fabric to reduce the traffic volume on the foot over bridges, (2) a guideline for the structural audit of the bridges, (3) replacement of old foot-over-bridges with new foot-over-bridges, (4) safety policy at the city level for pedestrians.

7 Conclusions

The foot-over-bridges in global cities require a comprehensive guideline for the safety of the citizens. This guideline could be stringent methods for the structural audit. This method can be in three parts: (1) geometric studies, (2) structural testing, and (3) vibration analysis. Each of these parts needs to be investigated to identify the present scenario of the foot-over-bridges. Geometric studies includes the details investigation of physical details through visual observations. This observation can be undertaken for all the damaged parts of the foot-over-bridge. Most of the foot-over-bridges are critical with the joineries that determine the failure of the structure. The structural testing can provide a scientific basis for failure analysis of the foot over bridges. In the case of the steel structures, the joinery can be studied by the use of a dye penetration test. Concrete strength can be studied by the use of rebound hammer test, ultrasonic pulse velocity test and, core cutting and compression test. Vibration analyses indicate the minimum permissible deflections in the members.

The case study had thus indicated methods for the study of the foot-over-bridges. In this case, studies the following observations are found; (1) the strength of concrete is not good at various locations, (2) the thickness of the steel is reduced to critical condition, (3) the connection of steel members are heavily damaged, and (4) vibration analysis have shown the failure of members. Based on this conclusion the foot-over-bridge was recommended to be demolished from the point of view of safety of the citizens. Thus, similar methods can be adapted for the analysis of damaged foot-over-bridges for making safer cities.

References

Caetano, E., Cunha, A., & Moutinho, C. (2010). Vandal loads and induced vibrations on a footbridge. *Journal of Bridge Engineering, 16*(3), 375–382. https://doi.org/10.1061/(ASCE)BE.1943-5592.0000154.

Chandrikka, V., Soundarya Lakshmi, B., Praveenkumar, M., Shanmugavel, S., & Sivasurya, B. (2019). Analysis and design of cold formed steel foot over bridge. *International Research Journal of Engineering and Technology (IRJET), 6*(3), 2941–2946.

De Luca, A., Bauer, M., Pinto, M., & Malsch, E. (2018). Vibration analysis of footbridges: An overview of the current practice. In *MATEC Web Conference* (Vol. 211, p. 10002).

Lelegeron, F., Moine, M., & Paultre, P. (2000). Studies on the vibration of pedestrian bridges. *IABSE Congress Report,* 657–664.

Maca, J. (2017). *Dynamic response of footbridges with tuned mass dampers*. In *IOP conference series: Materials science and engineering* (Vol. 236)

Modena, C., Tecchio, G., Pellegrino, C., da Porto, F., Zanini, M.A., & Doná, M. (2014). Retrofitting and refurbishment of existing road bridges. In *Maintenance and Safety of Aging Infrastructure, Structures and Infrastructure Book Series* (Vol. 10, pp. 469–533). CRC Press. http://dx.doi.org/10.1201/b17073-18

Oliveira, C. (2017). Fundamental frequencies of vibration of footbridges in Portugal: From in situ measurements to numerical modelling. *Shock and Vibration,* Article ID 925437, 22 p.

Parlewar, P. (2021a). Rehabilitation of historic Chancery building, Yangon, Myanmar. In: J. Delgado (Eds.), *Case studies in building rehabilitation*. Building pathology and rehabilitation (Vol. 13). Cham: Springer.

Parlewar, P. (2021b). Prince of Wales Museum: Revitalization of European painting store. In: J. Delgado (Eds.), *Case studies in building constructions*. Building Pathology and Rehabilitation (Vol. 15). Cham: Springer.

Parlewar, P. (2021c). Rehabilitation of K. G. road complex at New Delhi, India. In: J. Delgado (Eds.), *Case studies in building rehabilitation*. Building pathology and rehabilitation (Vol. 13). Cham: Springer.

Prashanth, T., & Gokulnath, V. (2018). Design and analysis of foot over bridge using STAAD Pro. *International Journal of Civil Engineering and Technology (IJCIET), 9*(1), 639–651.

Rajesh, S. (2017). Design of a steel foot over bridge in a railway station. *International Journal of Civil Engineering and Technology, 8*, 1533–1548.

Salamak, M., & Fross, K. (2016). Bridges in urban planning and architectural culture. *Procedia Engineering, 161*, 207–212.

Shukla, A., & Maiti, P. R. (2019). Retrofitting and rehabilitation of damage footbridge over Yamuna river. *International Journal of Recent Technology and Engineering (IJRTE), 9*(2), 4239–4246.

Zivanovic, S., Pavic, A., & Reynolds, P. (2005). Vibration serviceability of footbridges under human-induced excitation: A literature review. *Journal of Sound and Vibration, 279*(1–2), 1–74.

Chloride Penetration in Reinforced Concrete Structures: Case Study of Arvoredos Island

Carlos Eduardo Tino Balestra

Abstract Reinforced concrete structures present in the marine environment are subject to the chloride penetration in the different marine aggressive zones. These chlorides penetrate through the concrete cover layer and, when reach the reinforcement, create favorable conditions for the beginning of the corrosion process, leading to several problems as the progressive reduction of the mechanical strength of the reinforcement. At this point, considering the existence of port infrastructures in operation, playing a key role in trade relations between nations, this chapter presents a case study regarding chloride attacks on reinforced concrete structures present in different marine aggressive zones under natural degradation for more than 40 years. The objective of this chapter is to inform managers responsible for the maintenance and rehabilitation of these structures about the aggressiveness of chloride attacks in the context of reinforcement corrosion through a case study presentation, thus demonstrating that the lack of maintenance strategies and rehabilitation plans can lead to loss of functionality and safety of the structure, where total or partial collapses are likely to occur.

Keywords Reinforcement corrosion · Case study · Service life · Concrete structures · Marine environment

1 Introduction

From the service life perspective of reinforced concrete structures, the alkaline pH of the solution present in the concrete pores (pH close to 13) creates favorable conditions for the formation of a passive film, composed of iron oxides, which is capable of coating the reinforcement and chemically protect it against corrosion. In addition, under a physical point of view, the concrete cover layer (which delimits the concrete surface to the depth where the reinforcement is located) protects the bars from direct

C. E. T. Balestra (✉)
Department of Civil Engineering and Department of Environmental Sciences, Federal University of Technology and Western Parana State University, Toledo, Paraná, Brazil
e-mail: carlosbalestra@utfpr.edu.br

J. M. P. Q. Delgado (ed.), *Building Pathologies and Acoustic Performance*,
Building Pathology and Rehabilitation 18, https://doi.org/10.1007/978-3-030-71233-4_2

contact with aggressive agents present in the external environment, providing protective conditions to the reinforcement. These durability factors, combined with the well-known physical-chemical and mechanical properties of reinforced concrete, justify the preference for the wide use of this cementitious composite in the civil construction industry (Mehta and Monteiro 2006; Neville and Brooks 2010; Broomfield 2007).

Although the reinforcement remains physically and chemically protected inside the concrete, cover layer presents a network of interconnected pores with variable saturation degree, thus allowing the entry of aggressive agents from the external environment (carried by water) that can iterate with the concrete, leading to expansive reactions and cracking, or with the reinforcement itself, triggering a corrosive process. At this point, considering the marine environment, the chlorides present in the different marine aggressive zones, penetrate through the concrete pores network by mechanisms of capillary absorption and diffusion until reaching the reinforcements and triggering the corrosive process (Castro et al. 2001; Apostolopoulos et al. 2013).

At this point, it is necessary to clarify the marine aggressive zones the mechanisms of chloride penetration in concrete. So, first of all, the marine environment is subdivided into four main aggressive zones: submerged, tidal, splash and the marine atmosphere. In the submerged zone the structures (or their parts) remain completely submerged, that is, in permanent contact with the sea water, below the minimum levels reached by the tide. The tidal variation zone comprises the maximum and minimum limits reached by the tide. The splash zone is located just above the tidal zone, receiving the splash of the waves formed from the breaking of the waves and, finally, the marine atmosphere zone is that where the structures do not come into contact with the water, however, they are subject to the effect of a salt spray (François et al. 2013; Apostolopoulos 2009). Figure 1 present a perspective of the different marine aggressive zones.

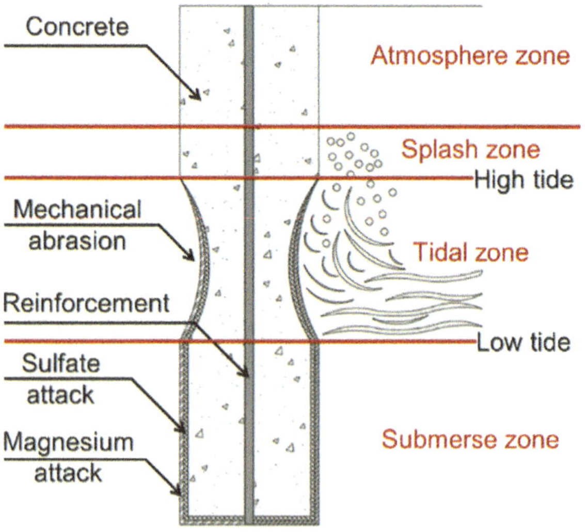

Fig. 1 Representation of marine aggressiveness zones. Adapted from (Mehta and Monteiro 2006)

From the perspective of reinforcement corrosion, tidal and splash zones are characterized as the most aggressive to reinforced concrete structures in the marine environment, since in addition to the significant presence of chlorides, there is a high availability of oxygen and moisture in the pores that are indispensable to the evolution of the corrosive process (Pang and Li 2016). At this point, it is worth mentioning that the structures present in these zones are subject, in a more intense way, to the wetting and drying cycles. In other words, during the period in which water comes into contact with the concrete surface (wetting period), it penetrates through the pore network due to capillary tensions acting on these pores (capillary absorption mechanism), allowing the chlorides transportation to the internal concrete layers, that is, the flow of water occurs from the external environment to the internal concrete layers. During the drying periods, this flow is inverted (from the internal concrete layers to the external medium), in which the water evaporates from the concrete pores leaving the chlorides inside the concrete. As this wetting and drying mechanism is cyclical, the chlorides concentration in the inner layers of the concrete increases progressively. Furthermore, considering the presence of moisture in the pores of the innermost layers of concrete, these chlorides from the external environment end up moving through the pore network due to ionic concentration gradients existing between layers closer to the surface (with high chloride concentrations and subject to wetting-drying cycles) to the innermost concrete layers (with a lower chloride concentration with pores constantly presenting some saturation degree), through the ionic diffusion mechanism governed by Fick's Laws of diffusion (Wang et al. 2018; Balestra et al. 2018).

Through these mechanisms mentioned above, chlorides end up entering the covering layer and when reaching the reinforcement in a concentration greater than 0.05% in relation to the concrete mass (Comité Euro-International Du Béton 1992), they end up destroying locally the passivating film, creating conditions to start the reinforcement corrosive process known as pitting corrosion (Fig. 2). The pits are characterized by well-defined points on the surface of the bars that deepen in an autocatalytic process, which may not lead to significant mass reductions in the bars, but significant damage to the mechanical properties of the reinforcements due to the cross-section reduction (Balestra 2016).

In the pitting corrosion process (represented in Fig. 3), the chlorides when in contact with the reinforcement lead to a locally acidify on the reinforcement surface destroying the passive film (Des-passivation). After des-passivation, chlorides remove the metallic from the bars through the formation of $FeCl_2$, which is dissociate in the pore concrete medium, allowing the combination of the metallic iron (Fe^{2+}) with hydroxyl ions (OH^-) present in the medium to the formation of iron hydroxides ($Fe(OH)_2$), known as corrosion products in the cathodic regions of the reinforcement. As the corrosion products have reduced mobility through the concrete pores, they are deposited in the reinforcement cathodic regions and, due to their greater volume in relation to the metallic iron, they end up exerting stresses on the concrete, leading to cracking and spalling of the cover layer. In addition, the chlorides (Cl^-), after being released, return to the internal and acidic region of the pit for new removals of metallic iron in a self-catalytic process. From this corrosion

Fig. 2 Reinforced concrete corrosion scheme due to chlorides action (Balestra et al. 2018)

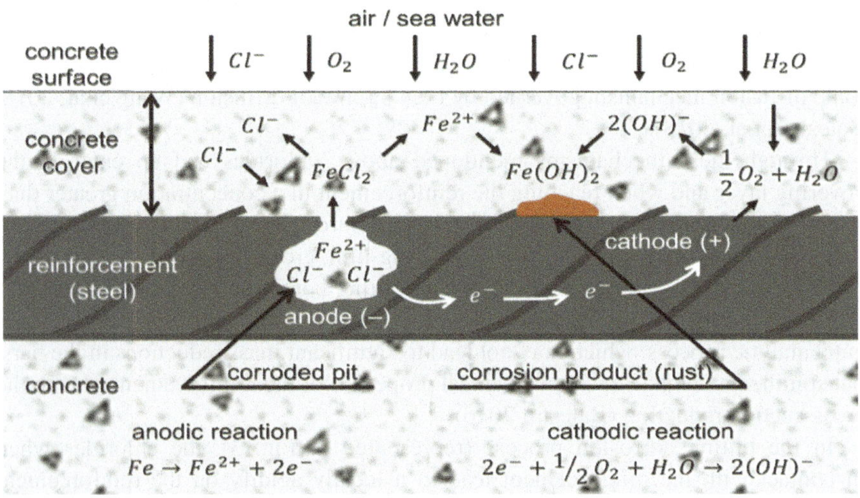

Fig. 3 Reinforced concrete corrosion scheme due to chlorides action (Mehta and Monteiro 2006; Neville and Brooks 2010; Broomfield 2007)

mechanism, a significant reduction of the active area of steel reinforcement can be observed, increasing the effective stress, and affecting the mechanical properties of the reinforcement (Mehta and Monteiro 2006; Neville and Brooks 2010; Broomfield 2007).

Based on the above, this chapter presents an analysis of reinforced concrete structures that remained under natural degradation in the marine environment for more than 40 years in different marine aggressive zones (marine atmosphere, splash and tidal). The objective is to present the real effects of chloride-induced corrosion on reinforced concrete structures, alerting managers responsible for the maintenance and rehabilitation of port infrastructures about the risks imposed by this pathological manifestation on these structures.

2 Arvoredos Island

The reinforced concrete structures analyzed in this chapter are present in Arvoredos Island (Coordinates: 23.967106 South (latitude) and 46.167061 West (longitude) This Island is a rock formation far from 1.6 km from the coast, located in the city of Guarujá, south coast of the São Paulo State, Brazil. The local climatic is classified as Cfa according to Köppen classification (Humid temperate climate with hot summer) (Peel et al. 2007). Figure 4 presents an overview of Arvoredos Island.

The history of the buildings at Arvoredos Island goes back to the 1950 s when the Island was given by the Brazilian navy to Eng. Fernando Lee for scientific research related to flora and fauna. For this purpose, several reinforced concrete structures were built on the Island between 1950 and 1974 in order to support the research carried out on the site. At this point, structures were built in different marine aggressive zones that have remained under natural degradation since its construction. In 2016, an extensive experimental program about the chloride penetration into these structures was conducted, including the development of a service life model through chloride profiles and analysis of surface electrical resistivity by 4-point method. The results obtained from this experimental program can consulted in the literature in References (Balestra et al. 2019a, b, c, d, e).

During the analysis of the structures, it was found that corrosion induced by chlorides was the main cause of degradation. At this point, some structures showed

Fig. 4 Overview of Arvoredos Island

Fig. 5 Degradation state of structures in Arvoredos Island: **a** Partial Collapse of a slab. **b** Column cross-section loss and intense reinforcement corrosion. **c** Micrograph showing pit corrosion

a severe degradation degree including partial collapses as shown in Fig. 5, where it is possible to notice the partial collapse of the slab of a structure (Fig. 5a) and the partial column section loss of a walkway with intense reinforcement corrosion (Fig. 5b). Furthermore, the micrographs obtained showed the presence of corrosion pits that justify the chlorides action (Fig. 5c).

The subsequent sub-chapters intend to present some structure analyzed at Arvoredos Island, highlighting the damages due to the reinforcement corrosion. It is worth mentioning that carbonation tests were performed on all of the structures analyzed, however the concrete carbonation was limited to a depth of less than few millimeters, indicating that corrosion attacks were exclusive due to chlorides action.

2.1 *Reservoir*

The reservoir is shown in Fig. 6. It is a concrete structure built in tidal zone with retaining walls of 1 m thickness approximately. This structure was built in order to develop researches about turtles. In this reservoir, flow and retention valves were installed to control the entry and exit of sea water to carry out the studies.

Since its conception, the degradation effects of a structure present in this aggressive marine zone were known. Because of this, it was decided to build a structure with very thick concrete walls (without reinforcements) in order to resist the sea waves impacts. By this way, this structure did not present reinforcements corrosion signs, being the pathologies observed referring to the mechanical effects of the waves shock

Fig. 6 Overview of reservoir structure (During low tide)

against the walls of the reservoir as shown in Fig. 6. In this structure, concrete cores were extracted to analyze the chloride concentrations; surface electrical resistivity test was performed during low tide and trace reconstitution were also performed in order to estimate the cement consumption. The results are presented in Table 1.

From the analysis of this structure, it was possible to observe that although the cement content was significant, the chloride concentration showed a value higher than the normative limit (0.05% in relation to the concrete mass). In addition, the electrical resistivity value observed is very low, indicating a high probability of reinforcement corrosion (if they existed in the structure).

At this point, in addition to the structure's location, in a marine aggressive zone that favors a high pore saturation degree and subject to wetting and drying cycles, another factor that contributed to this behavior was the use of large concrete aggregates (Fig. 7). The use of large aggregates favors the formation of an extensive transition zone at the interface of the aggregates with the cementitious matrix, whose porosity characteristics favored the ingress of water and chlorides, contributing for the results presented in Table 1. Thus, it is recommended to the designers responsible for the materials choice to be used in structures present in the tidal zone, take the size of the aggregates into consideration.

Table 1 Reservoir analysis results

Identification	Maximum chloride concentration (%, concrete)	Surface electrical resistivity (KΩ cm)	Cement content (kg/m^3)
Reservoir	1.043	1.30	439.94

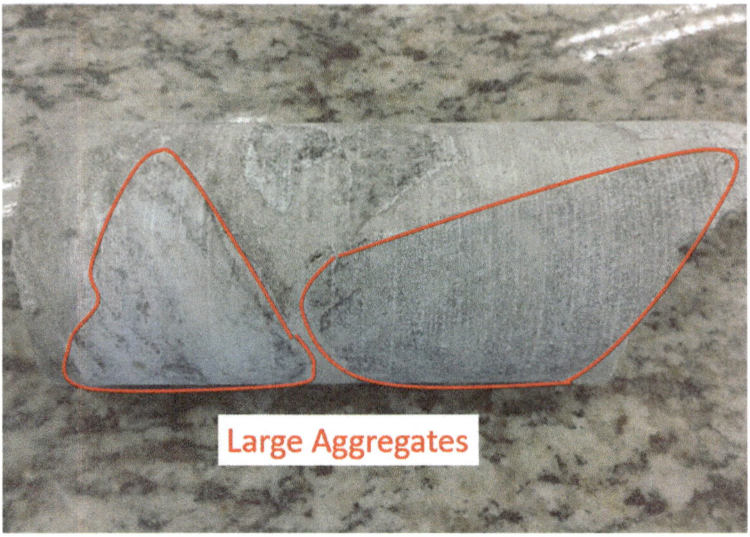

Fig. 7 Large aggregates observed in the concrete core from the reservoir structure

2.2 Platform and Baseline Structures

The platform and baseline structures are presented in Fig. 8. The platform has reinforced concrete elements built in the splash and marine atmosphere zones, whereas, baseline is a concrete element located in splash zone. Platform was a structure built to store small boats during unfavorable conditions to navigation. In this case, a crane was used to hoist the boats and reserve them in a pool as present in Fig. 9. Baseline is a concrete element with no structural function (and without reinforcement) that was used only for landing on the Island in the past.

Fig. 8 **a** Overview of platform structure. **b** Overview of baseline

Fig. 9 Storage of small boats in Platform

The Platform's structure present reinforced concrete columns and slab, where corrosion spots and cracks on surface attested to the advanced reinforcement corrosion degree of this structure. In this case, depending on the level reached by the tide, the lower parts of the three columns of the structure remained partially in splash or a marine atmosphere zone and, for this reason, the analyzes carried out were performed at the bottom part of these columns. Figure 10 present the degradation state of this structure, whereas, Table 2 present the parameters obtained from these structures.

The advanced degradation state of the platform structure is justified in view of the corrosion spots observed on the concrete surface (Fig. 10). In this case, these corrosion spots provide an indication of a high pore saturation degree, in such a way that, a portion of these corrosion products, instead of being deposited on the cathodic surface of the bars, these products were carried up to the surface of the structure, justifying including the low occurrence of concrete cracking. This fact also corroborates to the low values of surface electrical resistivity and the high concentration of chlorides observed in the analysis of this structure.

Regarding the level, although the baseline does not present reinforcement, the low values of superficial electrical resistivity combined with the concentration of chlorides, indicates the high aggressiveness to which structures present in this marine aggressiveness zone are subject. In this case, the contribution of the wetting and drying cycle is evident in the high chloride concentration observed.

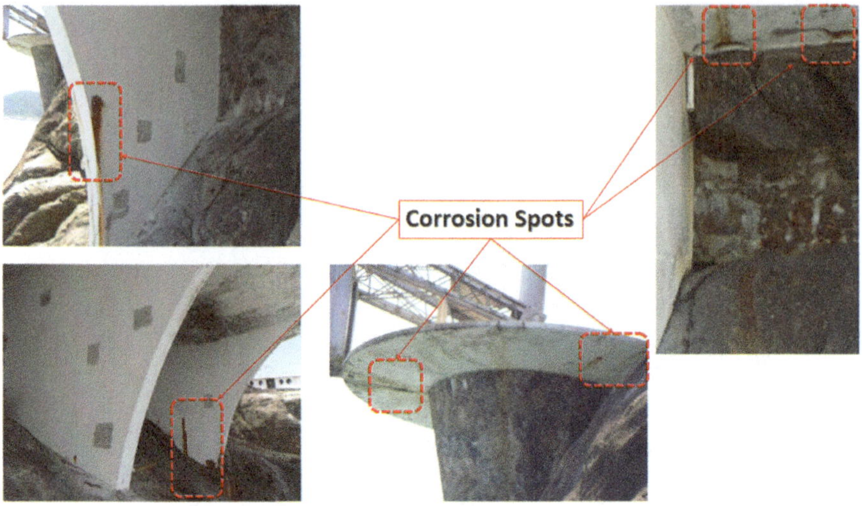

Fig. 10 Corrosion spots observed in platform structure

Table 2 Platform and Landing analysis results

Identification	Maximum Chloride concentration (%, concrete)	Surface electrical resistivity (KΩ cm)	Cement content (kg/m^3)
Platform	0.684	11.70	281.75
Baseline	1.867	2.50	–

2.3 *Stairs 1 and 2*

Bearing in mind that the Island is a rock formation, reinforced concrete stairs were built to access the different parts of the Island. In this case, stairway 1 has its lower steps located in a splash zone and stairway 2 is completely in a marine atmosphere zone. The aesthetics of both stairs followed the same pattern, with a central beam supporting the steps, however, the degradation degree due to corrosion is much greater in stairway 1, since the lower part of it is in a splash zone. Figure 11 presents an overview of both stairs, where it is even observed that the steps of stairway 1 collapsed, leaving only the central beam of the structure. Table 3 presents the data referring to these structures.

It is possible to note that in fact the splash zone is characterized as one of the most aggressive to reinforced concrete structures, in view of the advanced corrosion degree observed in stairway 1 due to the large availability of chlorides, moisture and oxygen in this zone. It is worth mentioning that stairway 1 is currently no longer used on the Island, in view of the risk of collapse of the central beam due to the reinforcement corrosion (Fig. 11a). This fact demonstrates in fact that the corrosion

Fig. 11 **a** Stairway 1 (Splash zone). **b** Stairway 2 (Marine atmosphere zone)

Table 3 Results of the analysis at Stairs 1 and 2

Identification	Maximum Chloride Concentration (%, concrete)	Surface electrical resistivity (KΩ cm)	Cement Content (kg/m^3)
Stair 1	1.451	2.30	–
Stair 2	1.088	17.50	137.11

problem is capable of making the use of a reinforced concrete structures unfeasible due to collapse risk.

Regarding the parameters presented in Table 3, it is possible to observe that both structures have low values of surface electrical resistivity, indicating a high corrosion probability as verified in visual inspection. In addition, the analyzes showed that the low cement consumption found in the stairway 2 structure corroborates for a concrete with high porosity, facilitating the chloride access from the external environment to the reinforcement region, even though the structure being present in marine atmosphere zone. Figure 12 presents the degradation state of stairway 2 were reinforcement corrosion can be observed.

Thus, in a general analysis, it is recommended that managers, designers and those responsible for the maintenance of port infrastructure pay attention to the concrete quality of these structures located in marine environment. At this point it is essential to conduct researches and design projects that target concrete with low porosity and permeability, since this is one of the key points to restrict the chlorides penetration into concrete and, concomitantly to that, increase the service life of reinforced concrete structures present in marine environment.

Fig. 12 Reinforcement corrosion in stairway 2

3 Conclusions

The marine environment is characterized as one of the most aggressive to reinforced concrete structures from the perspective of reinforcement corrosion. In this way, this work presented the analysis of structures under natural degradation for more than 40 years in different marine aggressive zones. The main conclusions are:

- Tidal and splash zones are the most aggressive to reinforced concrete structures under the reinforcement corrosion point of view. The severe degradation degree of the structures analyzed, including the interdiction of one of the structures analyzed due to the collapse risk combined with the low surface electrical resistivity observed, corroborate for this statement.
- Depending on the pore saturation degree, concrete cracking may not indicate the real damage imposed by the reinforcement corrosion, however, the evidence of corrosion spots on concrete surface are indicative of the severe degradation state of the structure and high degree of pore saturation.
- Concrete quality is one of the most important parameters that should be take into account when a structure present in marine environment will be built. For this, low porosity and permeability concretes are necessary for durable structures presente in marine environment.

References

Apostolopoulos, C. A. (2009). The influence of corrosion and cross-section diameter on the mechanical properties of B500c steel. *Journal of Materials Engineering and Performance, 18*(2), 190–195.

Apostolopoulos, C. A., Demis, S., & Papadakis, V. G. (2013). Chloride-induced corrosion of steel reinforcement: Mechanical performance and pit depth analysis. *Construction and Building Materials, 38*, 139–146.

Balestra, C. E. T. et al. (2016). Corrosion degree effect on nominal and effective strengths of reinforcement naturally corroded. *Journal of Materials in Civil Engineering, 28*(10), 04016103.

Balestra, C. E. T., Lima, M. G., Mendes, A. Z., & Medeiros-Junior, R. A. (2018). Effect of corrosion degree on mechanical properties of reinforcements buried for 60 years. *IBRACON Structures and Materials Journal, 11*(3), 474–498.

Balestra, C. E. T., Nakano, A. Y., Savaris, G., & Medeiros-Junior, R. A. (2019a). Reinforcement corrosion risk of marine concrete structures evaluated through electrical resistivity: Proposal of parameters based on field structures. *Ocean Engineering, 187*, 106167.

Balestra, C. E. T., Reichert, T. A., Pansera, W. A., & Savaris, G. (2019b). Chloride profile modeling contemplating the convection zone based on concrete structures present for more than 40 years in different marine aggressive zones. *Construction and Building Materials, 198*, 345–358.

Balestra, C. E. T., Reichert, T. A., & Savaris, G. (2019c). Contribution for durability studies based on chloride profiles analysis of real marine structures in different marine aggressive zones. *Construction and Building Materials, 206*, 140–150.

Balestra, C. E. T., Savaris, G., Nascimento, R. S., Medeiros Jr., R. A. (2019d). Relations between chloride determination in real concrete structures by X-ray Fluorescence Spectroscopy and by Potentiometric Titration. *Journal of Building Pathology and Rehabilitation, 4*(18).

Balestra, C. E. T., Reichert, T. A., Pansera, W. A., & Savaris, G. (2020). Evaluation of chloride ion penetration through concrete surface electrical resistivity of field naturally degraded structures present in marine environment. *Construction and Building Materials, 230*, 1169789.

Broomfield, J. P. (2007). *Corrosion of steel in concrete: Understanding, investigation and repair*. London: Taylor & Francis, 294 p.

Castro, P., de Rincon, O. T., & Pazini, E. J. (2001). Interpretation of chloride profile from concrete exposed to tropical marine environments. *Cement and Concrete Research, 31*, 529–537.

Comité Euro-International Du Béton. (1992). Bulletin D' Information N° 183: 1992: Durable concrete structures. 2.ed. Lausanne, SW: Telford Services, 122 p.

François, R., Khan, I., & Dang, V. H. (2013). Impact of corrosion on mechanical properties of steel embedded in 27-year-old corroded reinforced concrete beams. *Materials and Structures, 46*(6), 889–910.

Mehta, P. K., & Monteiro, P. J. M. (2006). *Concrete: Microstructure, properties and materials*. 3th edition New York: Mc-Graw Hill, 674p.

Neville, A. M., & Brooks, J. J. (2010). *Concrete technology*. 2th ed. London: Pearson, 448p.

Pang, L., & Li, Q. (2016). Service Life prediction of RC structures in marine environment using long term chloride ingress data: comparison between exposure trials and real structures surveys. *Construction and Building Materials, 113*, 979–987.

Peel, M. C., Fynlaysoon, B. L., & Mmahon, T. A. (2007). Update world map of Köppen-Geiger climate classification. *Hydrology and Earth System Sciences, 11*, 1633–1644.

Wang, Y., Wu, L., Wang, Y., Li, Q., & Xiao, Z. (2018). Prediction model of long-term chloride diffusion into plain concrete considering the effect of heterogeneity of materials exposed to marine tidal zone. *Construction and Building Materials, 1593*, 297–315.

Techniques to Assess the Corrosion Resistance and Corrosion Rate of the Steel Embedded in Concrete

P. Chandru, J. Karthikeyan, and C. Natarajan

Abstract Corrosion is the most inevitable issue in the reinforced concrete structures. Assessing the corrosion resistance and corrosion rate of the rebar becomes essential to predict the service life of the structure and also to take necessary precautionary measures. This chapter discusses about the various field and laboratory techniques available to assess the corrosion resistance and corrosion rate of the steel embedded in concrete. The various experimental techniques such as half-cell potential, impressed voltage, electrical resistivity, macro cell, AC impedance, and polarization resistance tests are discussed below.

Keywords Corrosion mechanism · Corrosion types · Assessment techniques

1 Introduction—Mechanism and Types of Corrosion

Corrosion of the rebar embedded in concrete is an electrochemical process. Corrosion usually occurs in a rate depending on an equilibrium between opposite electrochemical reactions. First is the anodic reaction, which involves oxidation of metal, releasing electrons into the metal. Next is the cathodic reaction, in which a solution species (often O_2 or H^+) undergoes reduction, removal of electrons from the metal. If both the reactions are in equilibrium, then there will be no net electron flow (no current flow). These anodic and cathodic reactions can occur within a metal or on two different metals which are electrically connected. In the reinforced concrete, the steel remains passive due to the high alkalinity (pH more than 12) of the concrete pore solution. The high alkalinity passivates the rebar by forming a thin (approximately 10 nm) ferric oxide film (γ-Fe_2O_3) on its surface. As long as the steel remains passive, the corrosion does not takes place. When the steel becomes depassivated,

P. Chandru (✉) · J. Karthikeyan · C. Natarajan
Department of Civil Engineering, National Institute of Technology, Tiruchirappalli, Tamil Nadu 620015, India
e-mail: 403117002@nitt.edu

J. Karthikeyan
e-mail: jk@nitt.edu

J. M. P. Q. Delgado (ed.), *Building Pathologies and Acoustic Performance*,
Building Pathology and Rehabilitation 18, https://doi.org/10.1007/978-3-030-71233-4_3

it starts corroding which results in rust formation, loss of bond at steel—concrete interface, delamination and spalling of cover concrete. The steel gets depassivated due to two major reasons such as carbonation of cover concrete and ingression of chloride ions.

In general, the carbonation is a chemical reaction through which the carbon dioxide produces carbonates, bicarbonates and carbonic acid. The carbonation process of the concrete can be explained using the Eqs. (1) and (2).

$$Ca(OH)_2 + CO_2 \rightarrow CaCO_3 + H_2O \quad (1)$$

$$CO_2 + H_2O \rightarrow H_2CO_3 \text{ (Carbonic acid)} \quad (2)$$

$$H_2CO_3 + Ca(OH)_2 \rightarrow CaCO_3 + 2H_2O \quad (3)$$

The atmospheric CO_2 reacts with the portlandite ($Ca(OH)_2$) present in the atmosphere to form calcium carbonate ($CaCO_3$). In other perspective, the atmospheric CO_2 reacts with the moisture and forms carbonic acid (H_2CO_3) on the surface of the concrete. Due to the depletion of portlandite and formation of carbonic acid, the pH value of the concrete gets decreased which results in depassivation of the steel. The carbonation process is relatively slower for the concrete exposed to very dry condition and fully saturated condition. But for the structures subjected to wetting and drying cycles, the carbonation process is relatively fast. Carbonation progresses rapidly at the relative humidity of 50–70%. The carbonation of concrete often leads to a generalized corrosion which can be identified through the formation of cracks in the direction parallel to the rebar. For the concrete structure with 25 mm cover, the cracks due to carbonation induced corrosion will begin only after 20 years of exposure. In the case of generalized corrosion, the loss of cross-section of the steel will be comparatively less. The carbonation process cause microcell corrosion in the rebar, whereas the chloride ingression causes macrocell corrosion (Alexander et al. 2017).

The ingress of chlorides into the concrete causes a localized corrosion called as pitting corrosion. Sea water and de-icing salts are the most common source of chlorides to the concrete. The ingressing chloride ion reaches the rebar, leading to depassivation followed by pitting corrosion. From Fig. 1 it can be observed that the ingress of chloride into the concrete causes a pitting corrosion, which makes the rebar to corrode only at a particular region. The pitting corrosion can cause more cross-sectional loss in the rebar. The stress corrosion cracking shown in Fig. 1 is caused by the unified action of stress and corrosion in the rebar. The stress corrosion cracking is the main reason for the failure of prestressed concrete structures. The prestressing tendons undergoes brittle failure due to the stress corrosion cracking (Vu et al. 2009).

According to the tutti's model shown in Fig. 2, service life of the reinforced concrete can be classified as two stages; one is the initiation stage and the next is propagation stage During the initiation stage, the carbonation and chloride ion

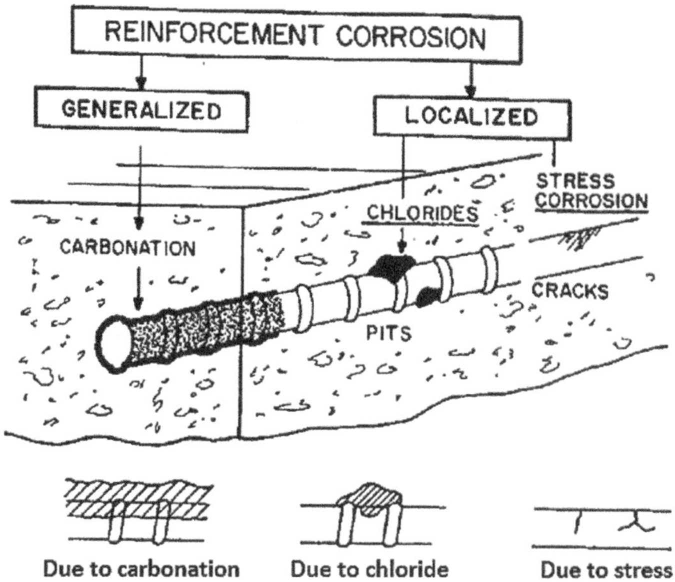

Fig. 1 Types of corrosion (Vavpetič 2008)

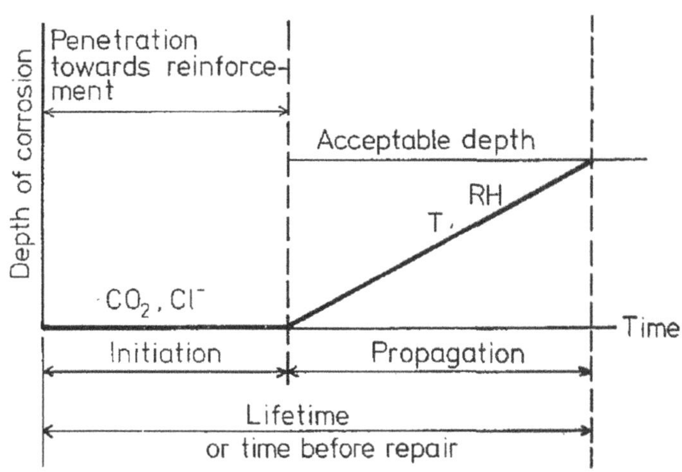

Fig. 2 Corrosion model for the steel embedded in concrete (Tutti 1982)

ingression occurs in the concrete. When the propagation stage begins, the steel gets fully depassivated and progresses to a point where no further corrosion is acceptable. The oxygen and moisture are necessarily to be present for the progress of corrosion, without which the corrosion progress becomes insignificant. Therefore, the presence of wider cracks in the concrete accelerates the corrosion process by allowing the

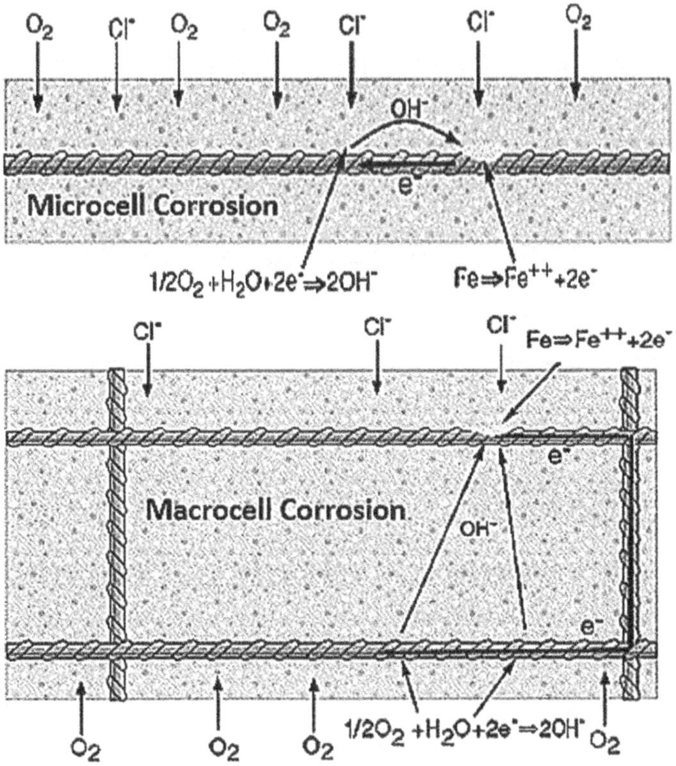

Fig. 3 Mechanism of microcell and macrocell corrosion (Hansson et al. 2006)

moisture, oxygen and chloride ions to easily reach the steel reinforcement (Suvash and Adewumi 2018) (Fig. 3).

The corrosion of rebar embedded in the concrete can be either microcell or macrocell. The Fig. 5 illustrates the difference between the microcell and macrocell corrosion. In the microcell corrosion, the anode and cathode will be very closer and this type of corrosion occurs due to the carbonation of concrete. In the case of macrocell corrosion, there will be a small localized anode and a large cathode. The chloride ingression causes macrocell corrosion with higher loss in the cross-section of steel and higher corrosion rates. Higher the chloride ion concentration destroys the passivating layer existing over the rebar surface. In order to prevent the chloride induced corrosion, the soluble chloride content should not exceed 0.15% for the reinforced concrete and 0.06% for the prestressed concrete (Poursaee 2016). From the Fig. 4, it can be observed that the corrosion of steel leads to the formation of rust ($Fe_2O_3.3H_2O$), which is around six times larger than the volume of the original metal. The increase in volume of metal by the formation of corrosion products leads to the loss of bond at steel—concrete interface, delamination and spalling of cover concrete.

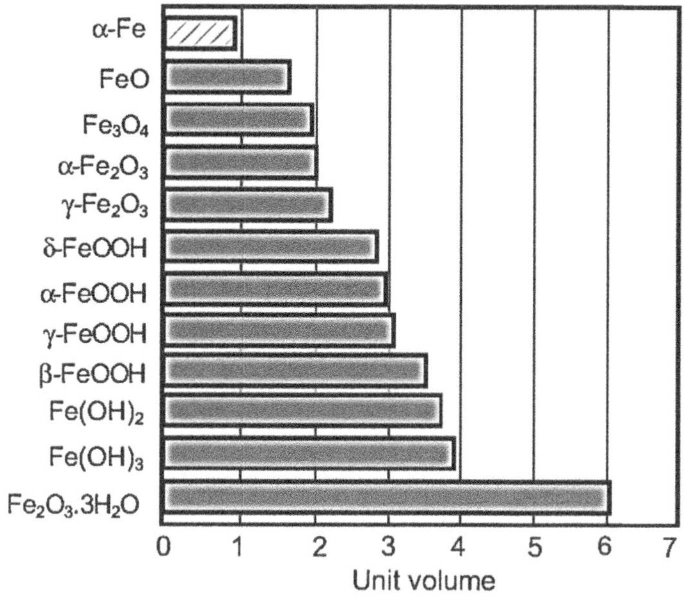

Fig. 4 Relative volume of the corrosion products (Poursaee 2016)

Corrosion of steel embedded in the concrete is an electrochemical process, it occurs due to the formation of cathodic and anodic zones on the surface of rebar as the result of electro-galvanic potential differences, with the concrete pore solution serves as an electrolyte (Alexander et al. 2017). The mechanism of carbonation induced corrosion can be understood using the following, Eqs. (4) to (8).

$$\text{At anode}: \text{Fe} \rightarrow \text{Fe}^{2+} + 2\text{e}^{-} \text{(oxidation reaction)} \tag{4}$$

$$\text{At cathode}: 2\text{H}_2\text{O} + \text{O}_2 + 4\text{e}^{-} \rightarrow 4\text{OH}^{-} \text{(reduction reaction)} \tag{5}$$

$$\text{Fe}^{2+} + 2\text{OH}^{-} \rightarrow \text{Fe(OH)}_2 \text{(ferrous hydroxide)} \tag{6}$$

$$4\text{Fe(OH)}_2 + \text{O}_2 + 2\text{H}_2\text{O} \rightarrow 4\text{Fe(OH)}_3 \text{(ferric hydroxide)} \tag{7}$$

$$2\text{Fe(OH)}_3\text{Fe}_2\text{O}_3 \rightarrow \text{H}_2\text{O} + 2\text{H}_2\text{O} \text{(Hydrated ferric oxide)} \tag{8}$$

The mechanism of chloride induced corrosion can be understood through the following Eqs. (9) to (13),

$$\text{At anode Fe} \rightarrow \text{Fe}^{2+} + 2\text{e}^{-} \text{(oxidation reaction)} \tag{9}$$

$$\text{At cathode}: \ H_2O + 1/2O_2 + 2e^- \rightarrow 2OH^- \ (\text{reduction reaction}) \tag{10}$$

$$Fe^{2+} + 2Cl^- \rightarrow FeCl_2 \ (\text{ferrous chloride}) \tag{11}$$

$$FeCl_2 + 2OH^- \rightarrow Fe(OH)_2 \ (\text{ferrous hydroxide}) + 2Cl^- \tag{12}$$

$$2Fe(OH)_2 + 1/2O_2 \rightarrow Fe_2O_3 \ (\text{ferric hydroxide}) + 2H_2O \tag{13}$$

$$4Fe(OH)_2 + O_2 \rightarrow 4\gamma - FeOOH + 2H_2O \tag{14}$$

2 Experimental Techniques

The various experimental techniques such as half-cell potential, impressed voltage, electrical resistivity, macro cell, AC impedance, and polarization resistance tests are discussed in the following sections.

2.1 *Half-Cell Potential Test*

The half-cell potential (HCP) method deals with the calculation of the electrical corrosion potential of uncoated rebar embedded in the concrete. The obtained corrosion potential is used to identify the corrosion activity in the rebar. This method can be used in field as well as in the laboratory. This test is done in accordance with ASTM C876 standard. From Fig. 5, it can be seen that the test setup consists a voltmeter, which measures the potential difference b/w the rebar and a copper sulfate ($CuSO_4$)/silver chloride (AgCl) reference electrode in contact with the concrete surface. The measured potential difference indicates the probability of corrosion in the rebar. During the test, the positive terminal of the voltmeter is connected to the rebar embedded in concrete and the negative terminal is connected with the reference electrode (AgCl/$CuSO_4$) which is placed over the concrete surface. The interpretation given in Table 1 provides the probability of corrosion in the rebar from the measured potential difference.

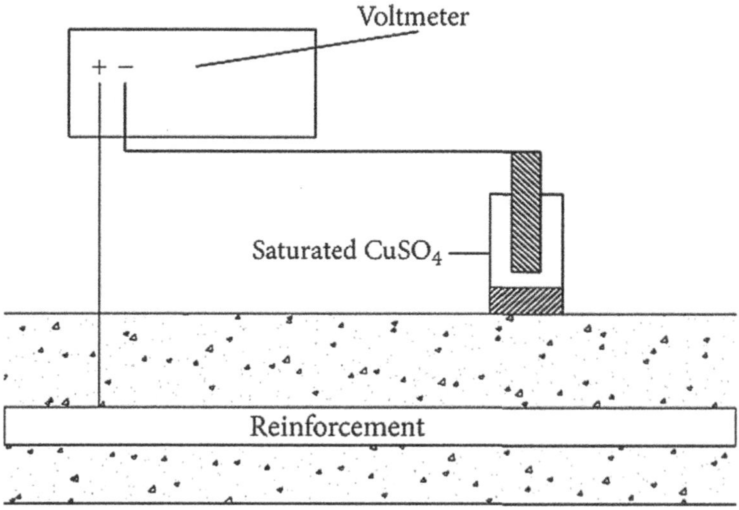

Fig. 5 Schematic view of half-cell potential test (Sadowski 2013)

Table 1 Interpretation for half-cell potential test

Corrosion potential (E_{corr}) values		Probability of corrosion
mV versus AgCl electrode	mV versus $CuSO_4$ electrode	
$E_{corr} > -125$	$E_{corr} > -200$	10% (negligible corrosion risk)
$-26 \leq E_{corr} \leq -275$	$-200 \leq E_{corr} \leq -350$	Intermediate corrosion risk
$E_{corr} < -276$	$E_{corr} < -350$	90% (high corrosion risk)
$E_{corr} < -426$	$E_{corr} < -500$	Severe corrosion

2.2 *Impressed Voltage Test*

The impressed voltage accelerated corrosion test (IVACT) is a laboratory technique to determine the corrosion resistance of the rebar embedded in the concrete. This test is usually conducted in the cylindrical concrete specimen with a rebar embedded in the centre. (Saraswathy et al. 2017; Rattanachu et al. 2020; Sharkawi and Seyam 2019; Wang et al. 2019). The schematic view of the experiment is shown in Fig. 6. The stainless-steel plate and the cylindrical concrete specimen are partially immersed in the NaCl solution (3–5% concentration). The positive terminal of the DC supply is connected to the embedded rebar (anode) and the negative terminal is connected to the stainless-steel plate (cathode). Then a constant DC supply of 5 to 15 V is applied at the anode. Then the corrosion current (mA) is continuously recorded at the anode. This test is continued up to the formation of cracks on the surface of the concrete.

The current flow in the specimen during IVACT can be described in three stages called the descending stage, steady stage, and ascending stage as shown in Fig. 7. In

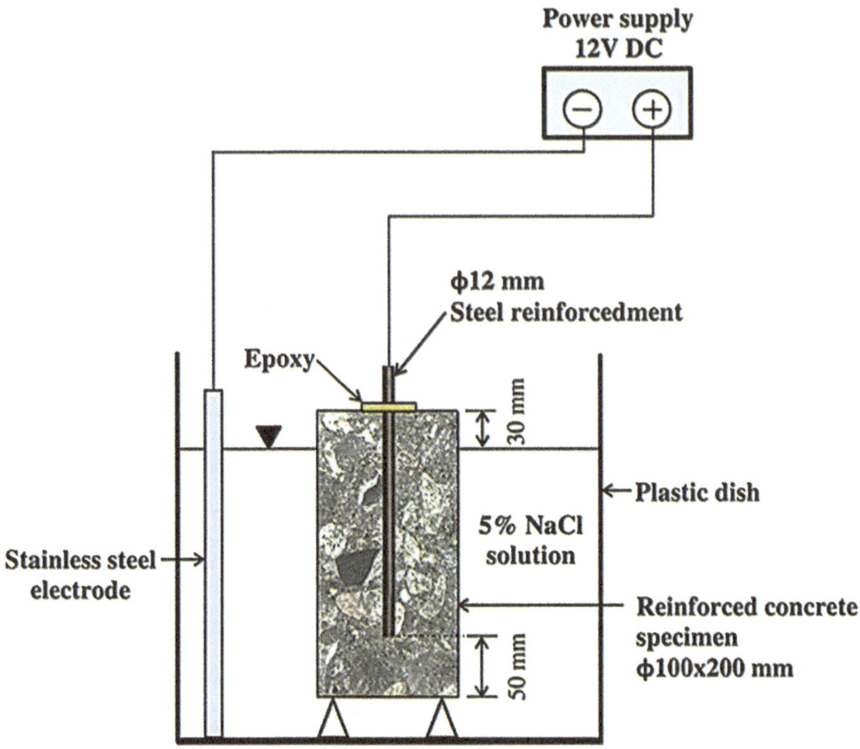

Fig. 6 Schematic view of IVACT (Rattanachu et al. 2020)

the descending stage, the corrosion products fill the interface between the embedded steel and concrete, which minimize the entry of gas as well as the flow of ions. So, the corrosion current gets reduced due to the inhibited anodic reaction. In the steady stage, equilibrium was attained between the ingression and consumption rate of the oxygen and moisture, which makes the current flow quite linear. Whereas in the third stage, the microcracks developed at the concrete-steel interface propagates and allow more oxygen and moisture to ingress. Therefore, the current flow gets gradually increased at this stage and further causes failure in the specimen (Wang et al. 2019). The Fig. 8 shows the typical failure mode of the concrete specimens subjected to the IVACT.

After the completion of test, the rebar embedded in the concrete specimen is removed, cleaned and weight loss in steel is measured. Then the corrosion rate is calculated using the given Eq. (15).

$$\text{Corrosion Rate} = 87.6 \times \left(\frac{W}{D \times A \times T} \right) \tag{15}$$

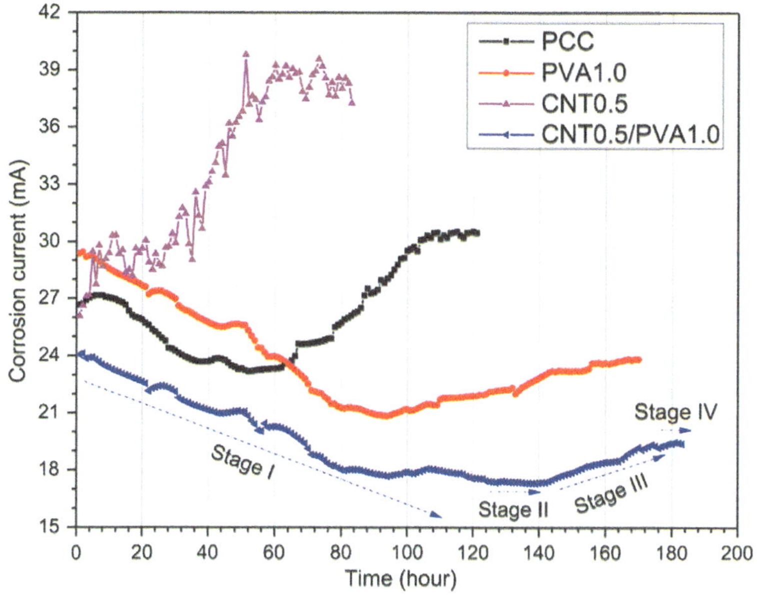

Fig. 7 Stages of corrosion in IVACT (Wang et al. 2019)

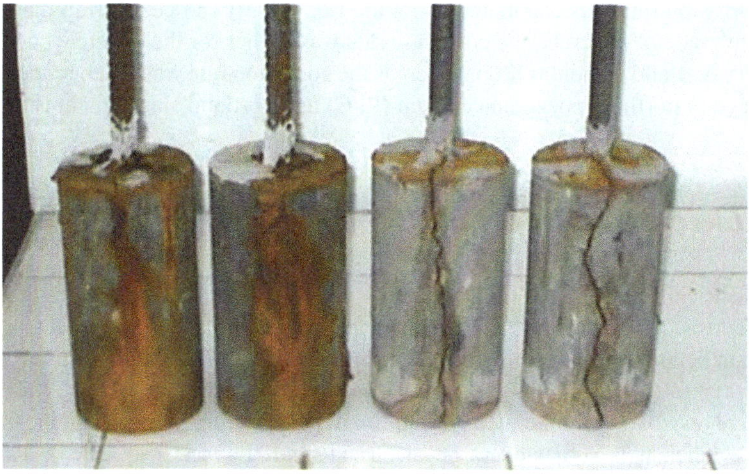

Fig. 8 Failure mode of specimens in IVACT (Güneyisi et al. 2005)

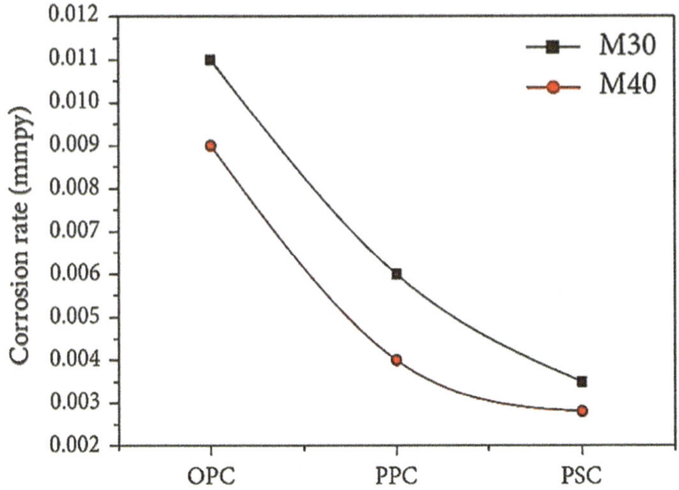

Fig. 9 Corrosion rates of the steel embedded in various cement concretes (Saraswathy et al. 2017)

In the Eq. (15), W = Mass loss in steel (mg); D = Density of steel (g/cc); A = Area of the specimen (cm^2); T = Duration of the IVACT (hours). The Fig. 9 shows the corrosion rate of the steel embedded in concrete containing different types of cement. The corrosion rate was calculated using the Eq. (15). It can be inferred that, in the both the grade of concrete, the corrosion rate was higher for the concrete containing ordinary portland cement (OPC). Whereas the corrosion rate was much reduced with the usage of portland pozzolana cement (PPC) and portland slag cement (PSC).

2.3 Electrical Resistivity Test

The electrical resistivity of the concrete is the parameter which is strongly related to the corrosion and chloride ion penetration resistance. Higher electrical resistivity results in better corrosion and chloride resistance; lower electrical resistivity results in high risk of corrosion and chloride permeation (Chandru et al. 2020, 2021). The electrical resistivity of concrete is the ability of the concrete to withstand the transfer of ions, when it is subjected to an electrical field (Layssi et al. 2015). This test is conducted in the saturated concrete specimen. Under saturated condition, in an applied electric field, the concrete with more interconnected pores transfer more ions and reduces the electrical resistivity of the concrete. Whereas in the concrete with low interconnected pores, the transfer of ions is negligible and results in better electrical resistivity. The test procedure is described in AASHTO T358, which is based on the Wenner four-probe principle. The Fig. 10 shows the schematic view of the electrical resistivity test on concrete. Table 2 presents the interpretation of concrete quality using the electrical resistivity values. This test can be carried out in the field as well

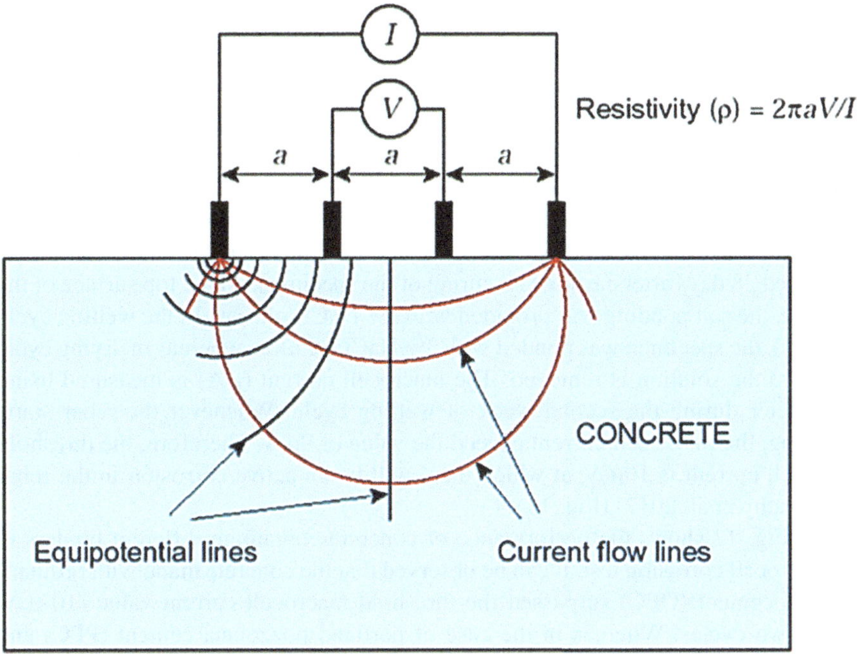

Fig. 10 Schematic view of electrical resistivity test on concrete (Yoon and Chang 2020)

Table 2 Interpretation for electrical resistivity test

Electrical resistivity (kΩ.cm) RILEM TC154—EMC		Electrical resistivity (kΩ.cm) AASHTO T358 & AASHTO TP95	
>20	Negligible corrosion	>254	Negligible Cl^- ion penetration
10–20	Low corrosion	37–254	Very low Cl^- ion Penetration
5–10	High corrosion	21–37	Low Cl^- ion Penetration
<5	Very high corrosion	12–21	Moderate Cl^- ion Penetration
		<12	High Cl^- ion Penetration

as in laboratory conditions. In the laboratory condition, this test is carried out using 100 mm × 200 mm cylinder specimens (saturated condition). During the test, a current is applied at the two outer probes, and the potential difference is measured b/w the two inner probes. The current is carried by the ions present in the pore solution of concrete.

2.4 Macrocell Test

The macrocell test is to determine the chemical induced corrosion resistance of the concrete. It is previously discussed that the ingression of chloride ions into the concrete causes a macrocell (anode and cathode are far away) type of corrosion. This test is described in the standard ASTM G109-13. The specimen details of this test are given in Fig. 10, from which it can be seen that the anode and cathode bars are away from each other. The ratio of anode to cathode area is 1:2. The test is conducted 28 days after casting and curing of the specimen. At the top surface of the concrete, the salt ponding was provided with 3% NaCl solution. In the wetting cycle (15 days), the specimen was ponded with 3% NaCl solution, whereas in drying cycle (15 days) the solution is removed. The macrocell current (μA) is measured using a voltmeter during the second week of wetting cycle. Whenever the rebar starts corroding, the macrocell current exceed the value of 9μA. Therefore, the threshold macrocell current is 10μA, at which there will be an active corrosion in the rebar (Saraswathy et al. 2017) (Fig. 11).

The Fig. 12 shows the performance of concrete containing different binders in the macrocell corrosion test. It can be observed that the concrete made with ordinary portland cement (OPC) surpassed the threshold macrocell current value (10 μA) within two cycles. Whereas in the case of portland pozzolana cement (PPC) and portland slag cement (PSC), it took almost 5 cycles to exceed the threshold macrocell current value (10 μA). This implies that the usage of fly ash/slag blended cement enhances the corrosion resistance of the concrete.

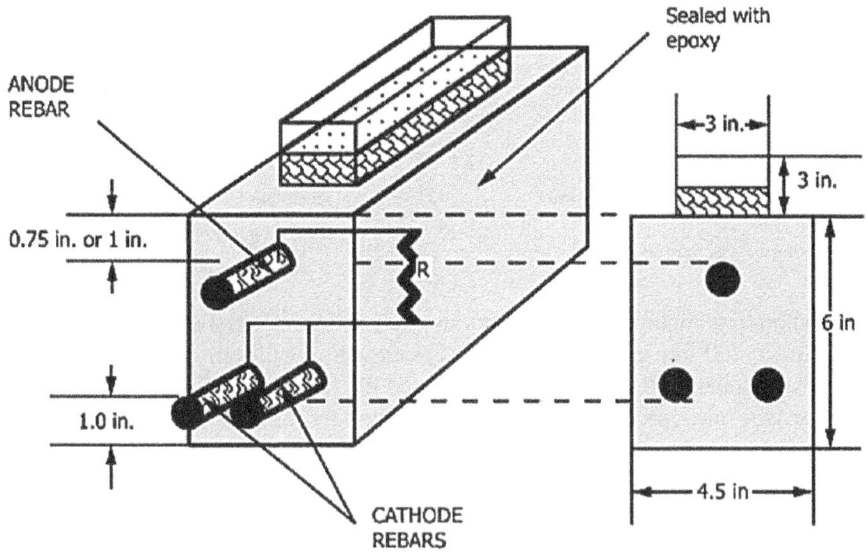

Fig. 11 Specimen details for macrocell corrosion test (ASTM G109-13)

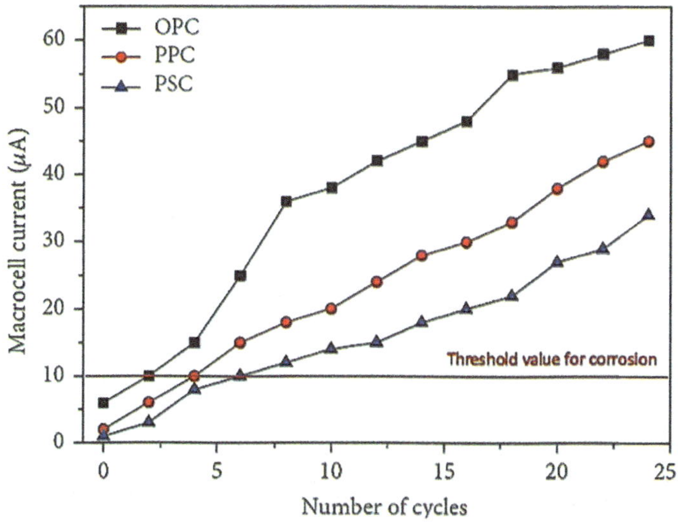

Fig. 12 Macrocell current measured from the steel embedded in various cement concretes (Saraswathy et al. 2017)

2.5 *Potentiodynamic Polarization Resistance Test*

In corrosion, the term polarization can be defined as a "potential shift" from its open circuit potential (zero-current potential) of the corroding system. If the potential shifts in a positive direction (above E_{corr}), then it is called as anodic polarization. If the potential shifts in the negative direction (below E_{corr}), then it is called as cathodic polarization. When the polarization is done potentiostatically, then the current is measured. If the polarization is done galvanostatically, then potential is measured. The potentiostatic is much adopted than the galvanostatic method. The potentiodynamic or cyclic polarization test requires three electrodes, an electrolyte and a potentiostat. In 3 electrodes, one is the working electrode (embedded rebar), second is an auxiallary electrode (which transports current to the working electrode and close the electrical circuit) and a reference electrode (which measures the potential difference). During the test, these three electrodes are immersed in an electrolyte solution and attached with the potentiostat. The potentiostat is used to apply potential to the electrical circuit and record the current induced in them. Since the potentiodynamic polarization test is mainly to assess the pitting corrosion, high anodic potential is applied to initiate the pitting corrosion (Fig. 13).

The corrosion parameters (E_{corr}, i_{corr}, β_a, and β_c) are estimated by intersecting the open circuit potential (E_{oc}) with the extrapolation of the linear portions of logarithmic current plot as shown in Fig. 14 (Daniyal and Akhtar 2020). The corrosion rate can be calculated by using the following Eq. (16).

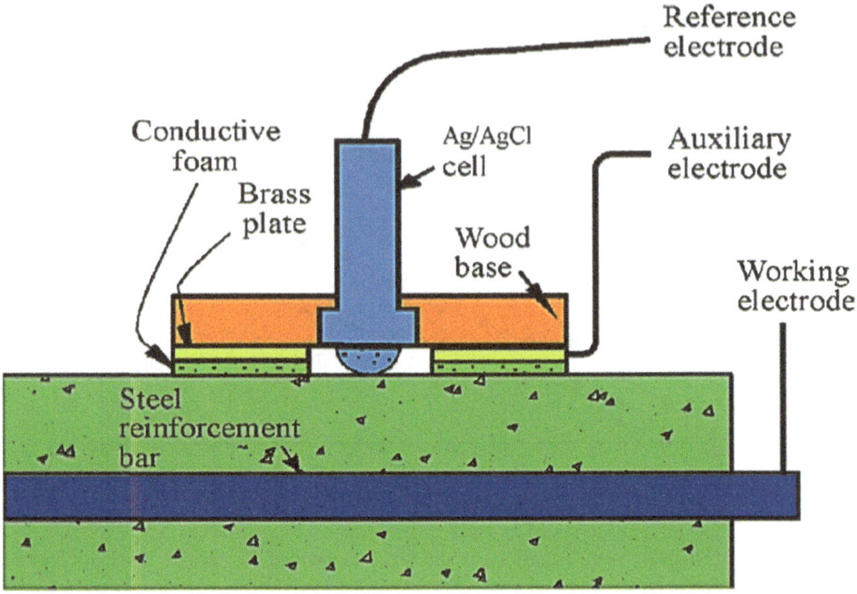

Fig. 13 Schematic view of polarization resistance test (Millard and Sadowski 2009)

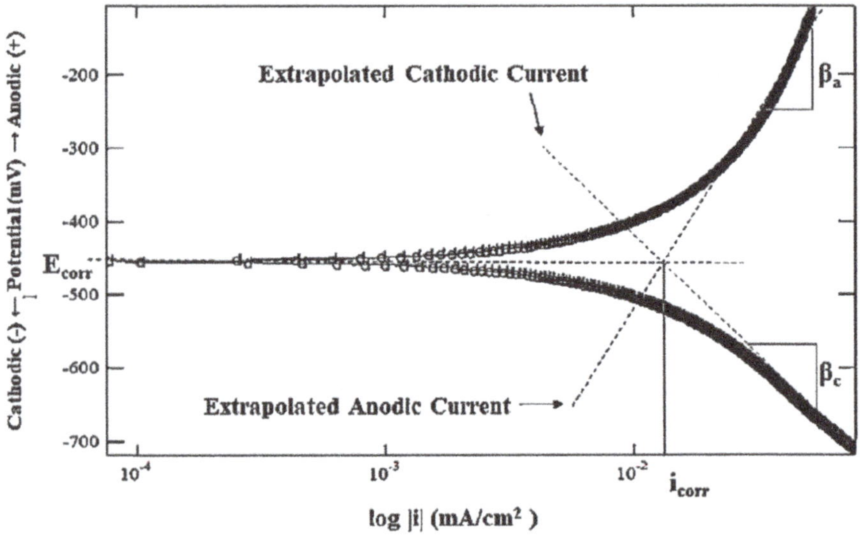

Fig. 14 Potentiodynamic polarization curve (Daniyal and Akhtar 2020)

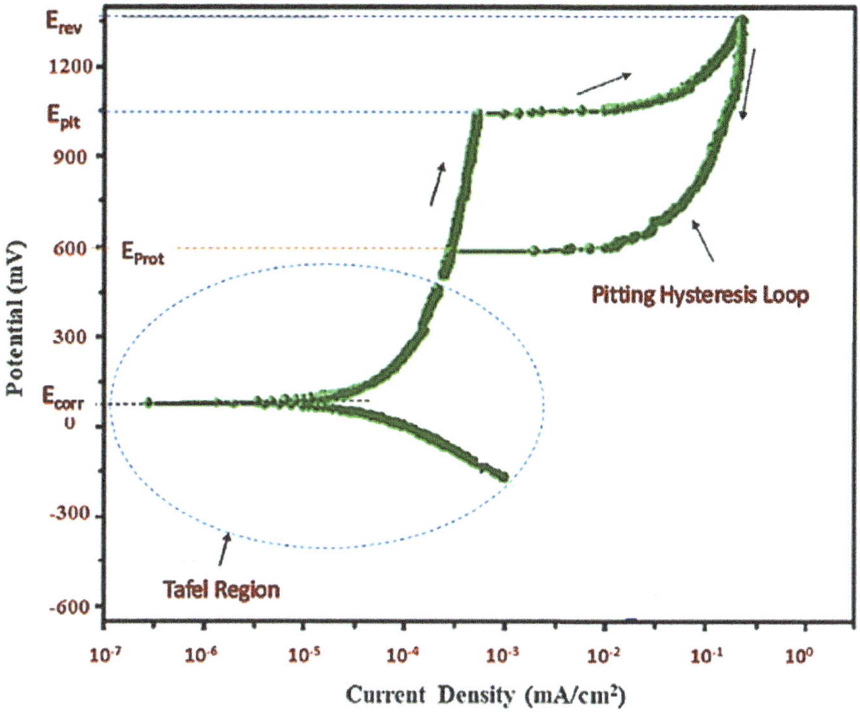

Fig. 15 Potentiodynamic polarization curve for pitting corrosion (Daniyal and Akhtar 2020)

$$\text{Corrosion Rate} = \frac{i_{Corr} \times K \times EW}{d \times A} \text{(mm/year)} \tag{16}$$

where EW (equivalent weight) = Theoretical mass of metal; i_{corr} = corrosion current density (mA/cm^2); K = 3272 (a constant that defines the units for the corrosion rate); d = density (g/cm^3); A = specimen area (cm^2). The Fig. 15 shows the typical cyclic polarization scan curve with the hysteresis loop, the larger hysteris loop indicates more severity in the pitting corrosion. In the shown scan plot Fig. 15, the term E_{rev} is the reverse potential, which is a predetermined anodic potential. Once the scan reaches the E_{rev}, it gets reversed towards the cathodic direction. E_{prot} is the protection potential, below which no pitting corrosion occurs. E_{pit} is the pitting potential beyond which there will be a sharp raise in the induced current. If E_{prot} is highly positive than the E_{pit}, then pitting corrosion does not takes place. If E_{prot} is highly negative than the E_{pit}, then pitting corrosion could takes place (Daniyal and Akhtar 2020).

The Fig. 16 shows the potentiodynamic polarization curve for the various cement concretes. From which it can be inferred that the corrosion risk is higher for the steel embedded in the ordinary portland cement (OPC) concrete when compared with the steel embedded in the portland pozzolana (PPC) and portland slag cement (PSC) concrete. In the steel embedded in OPC concrete, the corrosion current density

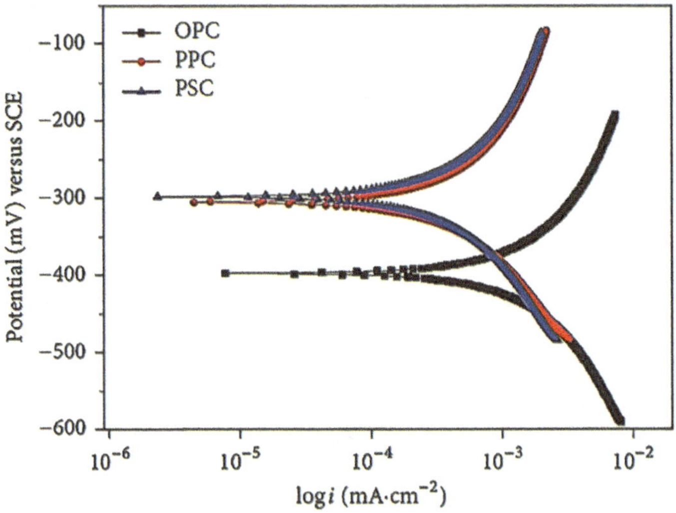

Fig. 16 Potentiodynamic polarization curve for the steel embedded in various cement concretes (Saraswathy et al. 2017)

Table 3 Interpretation for assessing corrosion risk in steel

Polarization resistance ($k\Omega.cm^2$) versus corrosion risk		Corrosion current density ($\mu A/cm^2$) versus corrosion risk		Corrosion penetration rate (μm/year) versus corrosion risk	
2.5–0.25	Very high	10–100	Very high	100–1000	Very high
25–2.5	High	1–10	High	10–100	High
250–25	Low/medium	0.1–1	Low/medium	1–10	Low/medium
>250	Passive	<0.1	Passive	<1	Passive

(mA/cm^2) is observed to be much higher and the open circuit potential (mV) is also more negative when compared with the steel embedded in PPC and PSC concretes. Table 3 provides the interpretation for assessing the corrosion risk using different parameters.

2.6 *Linear Polarization Resistance Test*

The Linear polarization resistance (LPR) is a corrosion assessment technique in which the corrosion rate of the steel embedded in concrete is assessed without causing much damage to the structure (more non-destructively as the applied potential is very low). In this method, the potential of around 25 mV is applied about E_{oc} using the potentiostat and the current induced is recorded. Since the applied potential is very

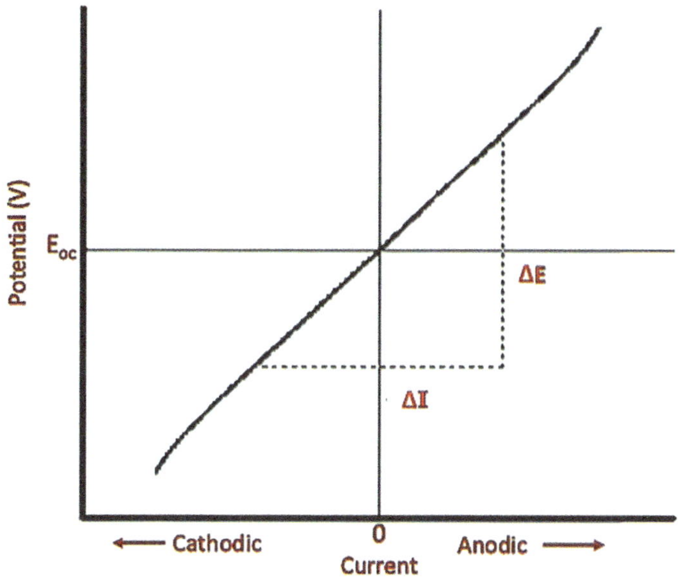

Fig. 17 Typical linear polarization curve (Daniyal and Akhtar 2020)

low, the induced current will be linear. The Fig. 17 shows the typical potential-current density (induced current/surface area of working electrode) graph obtained in LPR test (Daniyal and Akhtar 2020).

The linear polarization resistance (R_p) is obtained by taking slope at the potential-current density plot near E_{oc} as shown in Fig. 17. For the reinforced concrete structures, the linear polarization resistance (R_p) comprises of concrete resistance (R_c) and charge transfer resistance (R_{ct}). The linear polarization resistance (R_p) can be estimated using the Eq. (17).

$$\text{Linear resistance}\left(R_p\right) = \frac{\Delta E}{\Delta I} = R_c + R_{ct} \tag{17}$$

Then the Stern–Geary formula given in Eq. (18) can be used to determine the corrosion current density, In most of the instruments the value of B is 18 mV (Agrawal 2001).

$$i_{corr} = \frac{B}{Rp} \tag{18}$$

In the above Eq. (18), the i_{corr} refers to the density of the corrosion current; R_p refers to the linear polarization resistance (kΩ.cm^2) and B can be calculated using the following Eq. (19).

$$B = \frac{\beta_a \times \beta_c}{2.3(\beta_a + \beta_c) \times Rp} \tag{19}$$

In the Eq. (19), the β_a and β_c are the Tafel constants which can be determined as given in Sect. 2.5. The corrosion current density of more than 0.1 $\mu A/cm^2$ indicates the active corrosion in the steel embedded in concrete. Then the corrosion rate can be estimated by using the known i_{corr} values in the equation derived from the Faraday's law, as given in Eq. (20). In which, i_{corr} is the density of corrosion current; w is the weight of the steel (g); ρ is the density of the embedded rebar (g/cm^3). In the case of all the iron-based materials, the current density of 1 μAcm^{-2} corresponds to the corrosion rate of around 11 μm/year (Agrawal 2001).

$$\text{Corrosion Rate} = \frac{3.27 \times i_{corr} \times W}{\rho} (\mu m/year) \tag{20}$$

2.7 *AC Impedance Spectroscopy Test*

The term impedance can be defined as the ability of a circuit to resist the flow of electrical current. The alternating current (AC) impedance spectroscopy is also called as electrochemical impedance spectroscopy. The electrochemical impedance spectroscopy provides information about the electrical resistance and capacitance of the concrete (Suryanto et al. 2017). The capacitance of the material is the measure of its polarization of charges, whereas the resistance of the material is the measure of its ability to restrict the flow of ions. Measuring these electrical properties as the function of frequency under an applied electric field provides the area of AC impedance spectroscopy. The Fig. 18 shows the Nyquist plot and the simple equivalent circuit for the Nyquist plot. The complex impedance Z*(ω) of the material at an angular frequency (ω) under applied electrical field can be presented as given in Eq. (21).

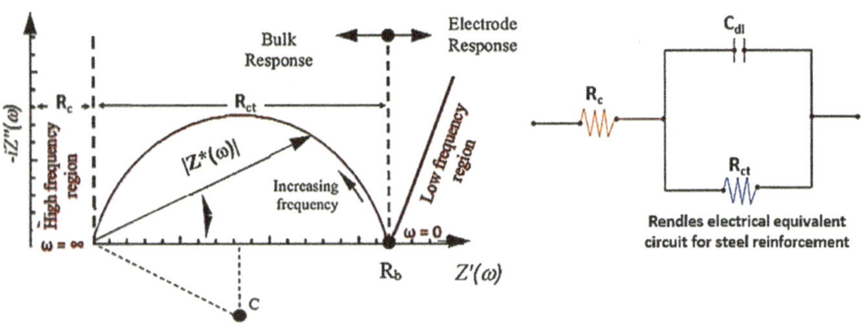

Fig. 18 Schematic diagram showing bulk response, electrode response and location of bulk resistance Rb (Suryanto et al. 2017; Daniyal and Akhtar 2020)

$$\text{Complex impedance } Z*(\omega) = Z'(\omega) - iZ''(\omega) \tag{21}$$

In the Eq. (21), the component Z'(ω) corresponds to the resistive phase and the component Z"(ω) corresponds to the reactive phase. In the Nyquist plot, both these components are plotted as the function of frequency under an applied electric field. It consists of a high frequency region (bulk arc) and a low frequency region (response from electrode). The R_b is the bulk resistance of the material.

The circuit equivalent to polarization resistance ($Z = R_c + R_{ct} = R_p$) is used in the AC impedance spectroscopy. The equivalent electrical circuit consists of two resistance (R_c and R_{ct}) and a double layer capacitor (C_{dl}). The R_c is the resistance of the cover concrete (resistance of the concrete pore solution). R_{ct} is the charge transfer resistance at the steel surface/steel-concrete interface (metal dissolution resistance). In this technique, a very low AC excitation potential of 5 to 20 mV (for obtaining linear response) with a varying frequency (100 kHz to 1 MHz) is applied to the rebar embedded in the concrete (working electrode) and the induced AC current response is recorded. The Fig. 19 shows the schematic diagram of AC impedance spectroscopy test. In the resulting induced AC current (linear response), the frequency will be identical to the applied AC potential (with different phase) with the difference in amplitude. As mentioned earlier, the recorded response (impedance) consists of two components Z'(ω) and iZ"(ω) which is analysed using a Nyquist plot.

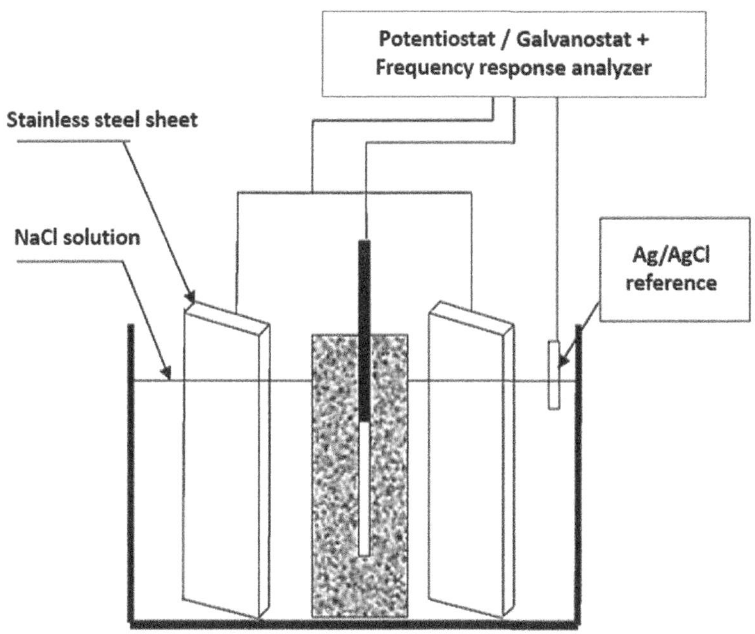

Fig. 19 Schematic diagram of AC impedance spectroscopy test (Ismail and Ohtsu 2005)

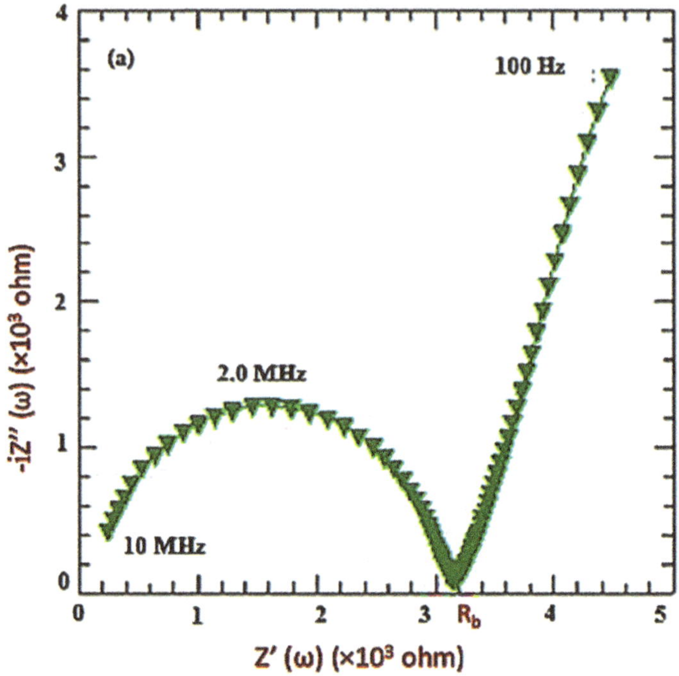

Fig. 20 Typical Nyquist curve for concrete (Suryanto et al. 2017)

The diameter of the semicircle shown in the Nyquist plot corresponds to the charge transfer resistance (R_{ct}) of the material (which is their true polarization resistance). If the diameter of the obtained semicircle is large, then it indicates that the material possesses a high polarization resistance (with low corrosion rate) and vice versa. The Fig. 20 shows the typical Nyquist curve for the concrete specimen tested in a frequency range of 100 Hz to 10 MHz.

3 Conclusions

From the results reported in the various studies, the following conclusions are provided,

- The steel gets depassivated due to two major reasons such as carbonation of cover concrete and ingression of chloride ions. In case of depassivation, it starts corroding which results in rust formation, loss of bond at steel—concrete interface, delamination and spalling of cover concrete.
- The carbonation process cause microcell corrosion in the rebar, whereas the ingress of chlorides into the concrete causes a localized corrosion called as pitting corrosion. In the microcell corrosion, the anode and cathode will be very closer.

In the case of macrocell corrosion, there will be a small localized anode and a large cathode. The macrocell corrosion leads to higher loss in the cross-section in steel and higher corrosion rates.

- The various experimental techniques such as half-cell potential, impressed voltage, electrical resistivity, macro cell, AC impedance, and polarization resistance tests are discussed, which are the effective techniques to assess the corrosion resistance of the concrete.

References

AASHTO T358. (2017). Standard Method of Test for Surface Resistivity Indication of Concrete's Ability to Resist Chloride Ion Penetration. American Association of State Highway and Transportation Officials, Washington, DC.

Agrawal, A. K. (2001). Corrosion Monitoring. 1698-1701. https://doi.org/10.1016/B0-08-043152-6/00302-8.

Alexander, M., Bentur, A., & Mindess, S. (2017). *Durability of concrete: Design and construction* (Vol. 20). CRC Press.

ASTM C876-15, Standard Test Method for Corrosion Potentials of Uncoated Reinforcing Steel in Concrete, ASTM International, West Conshohocken, PA, 2015.

ASTM G109-13, "Standard Test Method for Determining Effects of Chemical Admixtures on Corrosion of Embedded Steel Reinforcement in Concrete Exposed to Chloride Environments," in Proceedings of the ASTM International, West Conshohocken, PA, USA, 2013.

Chandru, P., Karthikeyan, J., & Natarajan, C. (2021). Correlations between the Hardened properties of combination Type SCC containing UFGGBFS. *Advances in Civil Engineering Materials, 10*(1), 1–22. https://doi.org/10.1520/ACEM20190233.

Chandru, P., Karthikeyan, J., Sahu, A. K., Sharma, K., & Natarajan, C. (2020). Some durability characteristics of ternary blended SCC containing crushed stone and induction furnace slag as coarse aggregate. *Construction and Building Materials*, 121483. https://doi.org/10.1016/j.conbuildmat.2020.121483.

Daniyal, M., & Akhtar, S. (2020). Corrosion assessment and control techniques for reinforced concrete structures: A review. *Journal of Building Pathology and Rehabilitation, 5*(1), 1. https://doi.org/10.1007/s41024-019-0067-3.

Güneyisi, E., Özturan, T., & Gesoğlu, M. (2005). A study on reinforcement corrosion and related properties of plain and blended cement concretes under different curing conditions. *Cement & Concrete Composites, 27*(4), 449–461. https://doi.org/10.1016/j.cemconcomp.2004.05.006.

Hansson, C. M., Poursaee, A., & Laurent, A. (2006). Macrocell and microcell corrosion of steel in ordinary Portland cement and high performance concretes. *Cement and Concrete Research, 36*(11), 2098–2102. https://doi.org/10.1016/j.cemconres.2006.07.005.

Ismail, M. A., & Ohtsu, M. (2005). Monitoring corrosion rate of concrete by AC impedance spectroscopy. In *11th International Conference on Fracture* (ICF XI) (pp. 20–25).

Layssi, H., Ghods, P., Alizadeh, A. R., & Salehi, M. (2015). Electrical resistivity of concrete. *Concrete International, 37*(5), 41–46.

Millard, S. G., & Sadowski, Ł. (2009). Novel method for linear polarisation resistance corrosion measurement. In *Proceedings of the NDTCE'09 Non-Destructive Testing in Civil Engineering*, Nantes, France, 30 June–3 July 2009.

Polder, R. B. (2001). Test methods for on site measurement of resistivity of concrete—a RILEM TC-154 technical recommendation. *Construction and Building Materials, 15*(2–3), 125–131. https://doi.org/10.1016/S0950-0618(00)00061-1.

Poursaee, A. (2016). Corrosion of steel in concrete structures. In *Corrosion of Steel in Concrete Structures* (pp. 19–33). Woodhead Publishing.

Rattanachu, P., Toolkasikorn, P., Tangchirapat, W., Chindaprasirt, P., & Jaturapitakkul, C. (2020). Performance of recycled aggregate concrete with rice husk ash as cement binder. *Cement & Concrete Composites, 108,* 103533. https://doi.org/10.1016/j.cemconcomp.2020.103533.

Sadowski, L. (2013). Methodology for assessing the probability of corrosion in concrete structures on the basis of half-cell potential and concrete resistivity measurements. *The Scientific world Journal*. https://doi.org/10.1155/2013/714501.

Saraswathy, V., Karthick, S., Lee, H. S., Kwon, S. J., & Yang, H. M. (2017). Comparative study of strength and corrosion resistant properties of plain and blended cement concrete types. *Advances in Materials Science and Engineering*. https://doi.org/10.1155/2017/9454982.

Sharkawi, A. M., & Seyam, A. M. (2019). Efficiency of accelerated techniques for assessing corrosion protection of blended cements. *Magazine of Concrete, 71*(12), 637–646. https://doi.org/10.1680/jmacr.17.00269.

Suryanto, B., McCarter, W. J., Starrs, G., & Chrisp, T. M. (2017). Characterization of fly-ash using electrochemical impedance spectroscopy. *Procedia Engineering, 171,* 705–714. https://doi.org/10.1016/j.proeng.2017.01.414.

Suvash, C. P., & Adewumi, J. B. (2018). A review on reinforcement corrosion mechanism and measurement methods in concrete. *Civil Engineering Research Journal, 5*(3), 555661. https://doi.org/10.19080/CERJ.2018.05.555661.

Tutti, K. (1982). Corrosion of steel in concrete. *Swedish Cement and Concrete Research Institute, of CBI research, 82–84,* 468.

Vavpetič, P. (2008). *Corrosion in concrete steel*.

Vu, N. A., Castel, A., & François, R. (2009). Effect of stress corrosion cracking on stress–strain response of steel wires used in prestressed concrete beams. *Corrosion Science, 51*(6), 1453–1459. https://doi.org/10.1016/j.corsci.2009.03.033.

Wang, Z., Yu, J., Li, G., Zhang, M., & Leung, C. K. (2019). Corrosion behavior of steel rebar embedded in hybrid CNTs-OH/polyvinyl alcohol modified concrete under accelerated chloride attack. *Cement and Concrete Composites, 100,* 120–129. https://doi.org/10.1016/j.cemconcomp.2019.02.013.

Yoon, I. S., & Chang, C. H. (2020). Effect of chloride on electrical resistivity in carbonated and non-carbonated concrete. *Applied Sciences, 10*(18), 6272. https://doi.org/10.3390/app10186272.

Influence of Mineral Additions in Improving the Chloride Binding Capacity of the Concrete

P. Chandru, J. Karthikeyan, and C. Natarajan

Abstract Ingression of chloride is a most devasting phenomenon in the concrete. Especially the concrete structures present at the coastal regions are highly vulnerable to the salt ingression and rebar corrosion. Therefore, it becomes essential to develop a durable cementitious composite which can withstand in the aggressive environments. Various researches and field experience have shown that the usage of blended cements can supress the ingression of chloride ions into the concrete. Blending the portland cement with the mineral additions like fly ash, ground granulated blast furnace slag, silica fume and metakaolin makes the concrete more impermeable and enhances its chloride resistance. The present chapter discusses about the possibility of improving the chloride resistance and chloride binding capacity of the concrete using various mineral additions.

Keywords Chloride resistance · Chloride binding · Fly ash · GGBFS · Silica fume · Metakaolin

1 Introduction

The ingress of chloride ion is the greatest threat to the rebar present in the reinforced concrete structure. The sea water and de-icing salts used on highways are the two major sources of chloride ions to the concrete. The chlorides in sea water are commonly found to be sodium chloride, magnesium chloride and calcium chloride. The salinity of the sea water is the major factor influencing the chloride concentration. For the sea water of salinity 35 g/L, the chloride concentration will be around 19,000 mg/L. The most commonly used de-icing salt is sodium chloride, whereas the magnesium and calcium chlorides are rarely used (Dyer 2014).

P. Chandru (✉) · J. Karthikeyan · C. Natarajan
Department of Civil Engineering, National Institute of Technology, Tiruchirappalli 620015, Tamil Nadu, India
e-mail: 403117002@nitt.edu

J. Karthikeyan
e-mail: jk@nitt.edu

J. M. P. Q. Delgado (ed.), *Building Pathologies and Acoustic Performance*,
Building Pathology and Rehabilitation 18, https://doi.org/10.1007/978-3-030-71233-4_4

When the chloride ions were ingressed into the reinforced concrete, they cause rebar corrosion which results in spalling of cover concrete and also impairs their mechanical properties. As shown in Fig. 1, after ingressing into the cementitious system, the chloride ions will be available in three forms which are chemically bound chloride (Friedel's salt and Kuzel's salt), physically adsorbed chloride (adsorbed within C–S–H layers), and free chloride ions (in pore solution). Among these three forms, the chemically bound chloride and physically adsorbed chloride has comparatively less potential to cause corrosion in rebar. Whereas the free chloride ions present in the cementitious system can easily migrate and cause rebar corrosion. Therefore, in order to prevent the chloride induced rebar corrosion in the reinforced concrete structures the following three remedial measures can be taken. (i) increasing the AFm phases in the cementitious system to chemically bind the ingressing chloride ions to form chloroaluminate complexes i.e., Friedel's salt and Kuzel's salt (Yang et al. 2020; Mesbah et al. 2012; Gbozee et al. 2018), (ii) promoting more C–S–H gel formation to physically adsorb the ingressing chloride ions (Liu et al. 2020a), (iii) refining the pore structure of the cementitious matrix to make them more impermeable to the gases and aggressive liquids (Chandru et al. 2020a).

The usage of mineral additions can effectively perform the above-mentioned remedial measures. Since the fly ash and GGBFS has a significant amount of Al_2O_3 as their chief chemical constituent, they can promote the formation of more Friedel's salt by associating with the ingressing chloride ions (Ma et al. 2018; Li et al. 2020). On the other side, the silica fume and metakaolin forms the dense C–S–H gel through the secondary hydration reaction which can effectively inhibit the ingression of chloride ions and also can physically adsorb the chloride ions within the interlayers of C–S–H gel. The various studies carried out to evaluate the chloride resistance and chloride binding capacity of concrete containing mineral additions are discussed in the following sections.

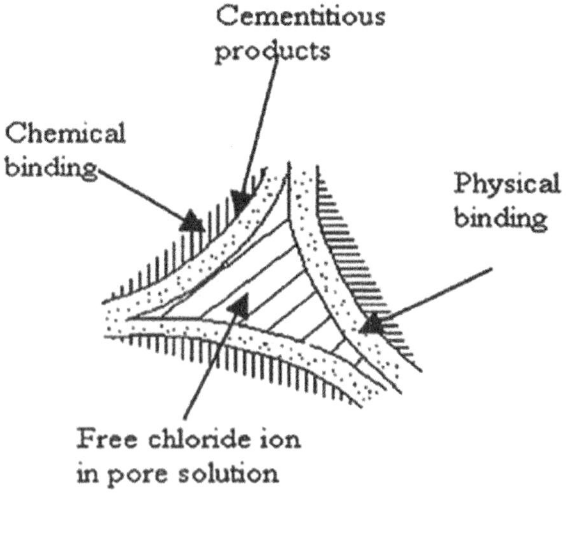

Fig. 1 Forms of chloride in concrete (Plang-ngern and Tangtermsirikul 1999)

2 Chloride Ingress in Concrete

The ingress of chloride ions into the concrete occur through permeation, diffusion, migration and sorption mechanism. The rate of chloride ingress into the concrete depends upon the type of mechanism, porosity of the concrete, constrictivity and tortuosity of cementitious system, chloride binding capacity of the cement matrix, and severity of the exposure condition at concrete surface.

2.1 Permeation

In permeation, the chloride ions ingress into the cementitious system due to a pressure differential. The flow rates of aggressive fluids through the concrete pores are sufficiently low for the flow to be categorized under laminar, therefore it can be described using Darcy's law given in Eq. (1).

$$Q = kA\left[\frac{(p_1 - p_2)}{\mu L}\right] \tag{1}$$

whereas, Q is the discharge (m^3/s); k is the permeability (m^2); A is the cross-sectional area through which the fluid is ingressing (m^2); $p_1 - p_2$ is the pressure differential (Pa); μ is the viscosity of the ingressing fluid (Pa s) and L is the distance between the pressure differential zones (m). Since the permeability k depends on the pore structure of the cementitious system and the properties of the ingressing fluid, the Eq. (2) can be used for calculating this parameter.

$$k = \frac{\left(r_e - P^2\right)}{S_e} \tag{2}$$

whereas, r_e is the effective pore radius (m); P is the effective porosity and S_e is the effective internal surface area of the porous material (m^{-1}). Moreover, the cracks existing in the cementitious system also promotes the ingression of fluids. Therefore, the permeability through the crack can be calculated using the Eq. (3).

$$k_{cr} = \varepsilon / w \tag{3}$$

whereas, k_{cr} is the permeability through crack; ε is the roughness coefficient; *l* and *w* are the length and width of the cracks (Meng 1994; Dyer 2014).

2.2 *Diffusion*

In diffusion, the chloride ions ingress into the cementitious system due to a concentration differential. The chloride ions diffuse through the pores existing in the cementitious system from the region of higher concentration (concrete surface) to the region of lower concentration (internal zones). The diffusion rate of the chloride ions increases with the increase in saturation level of the pores i.e., chloride ion diffusion occurs more in fully saturated pores than the partially saturated pores. The diffusion of chloride ions into the concrete under a concentration gradient can be described through Fick's law given in Eq. (4).

$$J = -D\frac{\partial c}{\partial x} \tag{4}$$

whereas, J is the transfer rate of the substance/unit area; $\partial c/\partial$ x is the concentration gradient; D is the diffusivity (m^2/s). The chloride diffusion coefficient values may range between 10^{-11} to 10^{-14} m^2/s. If the diffusion of chloride ions into the concrete becomes non-stationary, then it can be described through Fick's second law as given in Eq. (5). For non-stationary conditions, the chloride diffusion coefficient ranges between 10^{-11} to 10^{-13} m^2/s.

$$\frac{\partial c}{\partial t} = D\frac{\partial^2 c}{\partial x^2} \tag{5}$$

The diffusion of chloride is highly influenced by the pore structure of the concrete like volume permeable pores (effective porosity), constrictivity (narrowness of the pores along their length) and tortuosity (actual pore length/ shortest distance between the pore length calculated by drawing a straight line). Therefore, the effective chloride diffusion coefficient of the concrete can also be described using the Eq. (6) (Van and Heertjes 1974; Dyer 2014).

$$\mathrm{D_{eff}} = \varepsilon\frac{\delta}{\tau^2}\mathrm{D_s} \tag{6}$$

whereas, D_{eff} is the effective diffusion coefficient (m^2/s); ε is the pore volume; δ is the constrictivity of the pores; τ is the tortuosity of the pores and D_s is the self-diffusion coefficient of the diffusing fluid (m^2/s) (Currie 1960; Dyer 2014).

$$\delta = \frac{(c_{\mathrm{max}} \times c_{\mathrm{min}})^{1/2}}{c_{\mathrm{mean}}} \tag{7}$$

$$\tau = \frac{l'}{l} \tag{8}$$

The parameters constrictivity (δ) and tortuosity (τ) can be calculated from the above Eqs. (7) and (8) respectively. In which, C_{max}, C_{min} and C_{mean} are the maximum, minimum and average cross sectional are of the pore (m^2); l' is the average pore length (m) and l is the shortest distance between the pore length.

The typical chloride diffusion in concrete can be seen in Fig. 2. From the Fig. 2 it can be seen that the concentration of the chloride ion is higher at the surface of the concrete and lower at the interior sections. This is because, the diffusing chloride ions gets adsorbed at the surface of the concrete and increases the chloride ion concentration at the external sections. Due to the higher concentration of chloride ions at the external sections and clogging of pores due to the ingressing chloride salts, the diffusing rate of the chloride ions into the interior sections of the concrete gets significantly reduced. This phenomenon can be seen in Fig. 3

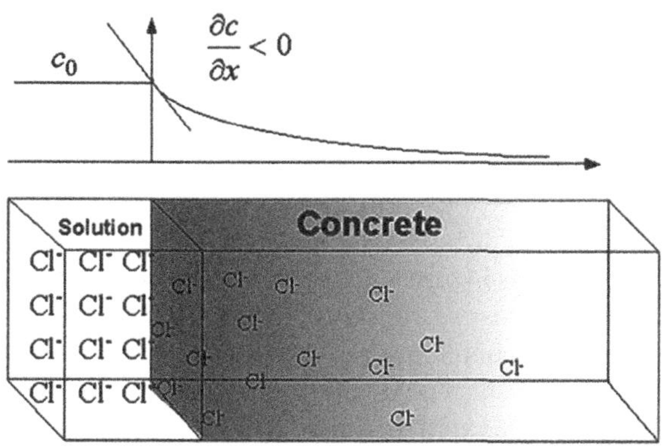

Fig. 2 Chloride diffusion in concrete (Tang et al. 2011)

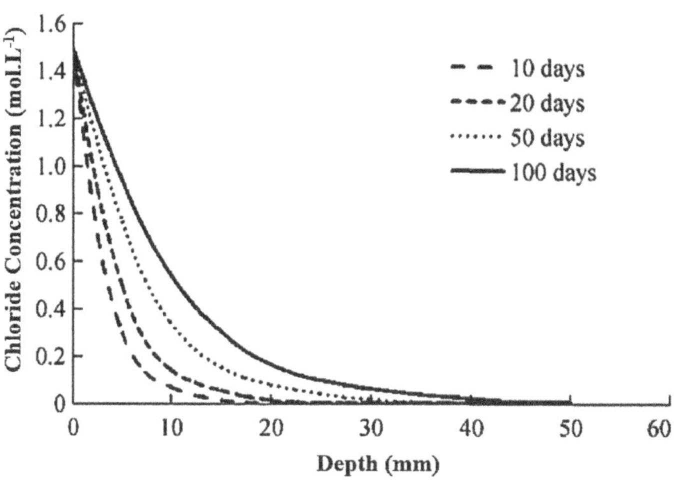

Fig. 3 Chloride concentration at various concrete depths (Damrongwiriyanupap et al. 2011)

2.3 Migration

In the case of migration, the chloride ions ingress into the cementitious system due to an electrical potential differential. Similar to the diffusion, the free chloride ions existing in the solution undergoes migration. The chloride diffusion coefficients of the concrete can be calculated using the Nernst-Einstein equation given in Eq. (9).

$$\mathrm{D_{eff}} = \frac{J_0 R T}{z F E_{c_0}} \tag{9}$$

whereas D is the chloride diffusion coefficient (cm^2/s); J is the flux of chloride ions (mol/cm^2s); R is the gas constant (8.314 J/K mol); T is the absolute temperature (300 K); L is the thickness of the specimen (cm); z is the valency of chloride ion (z = 1); F is the Faradays constant (9.648 × 104 J/V mol); C_0 is the initial chloride ion concentration (mol/l); and E is the potential (V).

2.4 Sorption

In sorption, the chloride ions ingress into the cementitious system through the capillary action of capillaries existing in the cement matrix. Capillary absorption is a dual-stage process such as initial and secondary, which depends on the interconnected pores existing in the concrete. In initial absorption ($0<\sqrt{t}<\sqrt{t_s}$), the fluid gets filled through absorption by capillary pores. In secondary absorption ($\sqrt{t} \geq \sqrt{t_s}$), the fluid saturation occurs by the diffusion and dissolution of air existing in the pore structure. Where 't' and 't_s' are the time and time to reach saturation (Fagerlund 1982; Esquinas et al. 2018; Chandru et al. 2020a). The adsorption/absorption of liquid into empty capillary pores can be expressed by the following relation Eq. (10).

$$\mathrm{i} = \mathrm{S}\sqrt{t} \tag{10}$$

where i is the depth of penetration; S is sorptivity and t is time. Sorptivity action will be stronger in the narrow capillary pores than in the wider capillary pores. But in narrow capillary pores, the rate of absorption become slower due to the higher friction.

3 Influence of Mineral Additions on Chloride Binding

The chloride binding capacity of the cementitious system is majorly influenced by the w/b ratio, degree of hydration, amount of aluminate phases (AFm) available and quantity of C–S–H gel produced. In the above-mentioned parameters, the aluminate phases present in the cementitious system binds with the ingressing chloride ion and forms chloroaluminate complexes (Friedel's salt) and the C–S–H gels adsorbs the ingressing chloride ions. Therefore, it is clear that increasing the availability of aluminate phases and C–S–H gel is the key to improve the chloride binding capacity of the concrete. The chloride binding capacity of the concrete containing different mineral admixtures are discussed below.

3.1 Fly Ash

Fly ash is a powder by-product discarded from the thermal power plants. It is a pozzolanic material consists of calcium, alumina and silica as a chief constituent. It is well known that the usage of fly ash as a part of binder enhances the mechanical and durability properties of the concrete. In the case of chloride binding also the performance of fly ash concrete was found to be superior to the conventional portland cement concrete. Irrespective of the type of fly ash used, the chloride binding capacity of the concrete increases with the increase in fly ash content. The SiO_2/CaO ratio plays a crucial role in the performance of fly ash based cementitious system. The chloride binding capacity of the fly ash based cementitious system gets increased till the value of SiO_2/CaO reaches 0.8. When $SiO_2/CaO > 0.8$, the chloride binding capacity of the cementitious system gets decreased, which was due to non-availability of sufficient $Ca(OH)_2$ to react with the fly ash for the production of secondary C–S–H gel. Increasing the specific surface area of the raw materials also have improved the chloride binding capacity of the cementitious system (Plang-ngern and Tangtermsirikul 1999). The binding of chloride in a cementitious system depends on two mechanism i.e. physical adsorption with C–S–H gel and chemical binding with AFm phases. The C_3A content of the binder has a good impact on the chloride binding capacity of the concrete. When comparing the chemical constituent of Type I portland cement with low/high calcium fly ash, it can be noticed that the Al_2O_3 content of portland cement will be around 5 to 7% and for fly ash it will be more than 20%. Due to the higher Al_2O_3 content, the fly ash based cementitious system can chemically bind greater amount of chloride ions when compared to portland cement based cementitious system.

The Fig. 4 shows that the increase in fly ash content decreased the availability of free chloride in the pore solution. The chloride ion diffusion coefficient was also decreased by 26% and 38%, when the portland cement was blended with 15% and 30% fly ash respectively. Moreover the chloride binding capacity index of the concrete containing 0%, 15% and 30% fly ash was found to be 0.1426, 0.1882 and 0.2134 respectively (Liu et al. 2020b). Apart from the chemical binding, physical adsorption of chloride ions by the hydration products also occurs. The ingressing chloride ions gets adsorbed with the hydration products like C–S–H, C–A–H, ettringite, and monosulfate (Cheewaket et al. 2010). The quantity of gypsum added to the cement also have a considerable influence in the chloride binding capacity of the

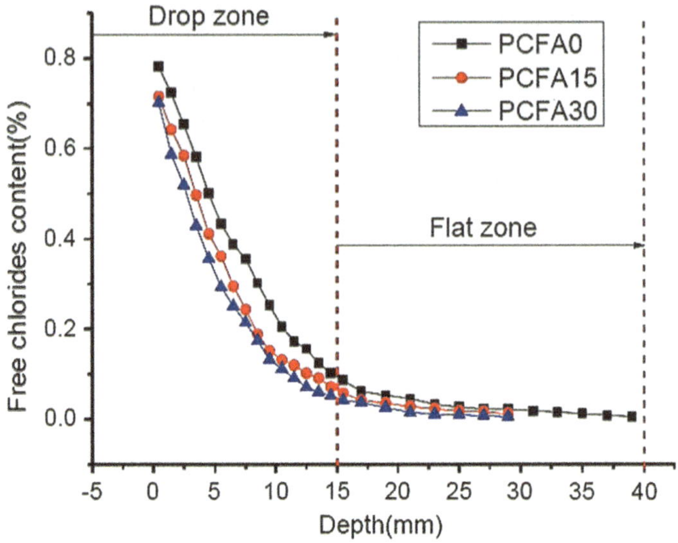

Fig. 4 Free chloride content at various depths of fly ash based concrete (Liu et al. 2020a)

cement paste. If the gypsum is overdosed, then it affects the formation of monosulfate and also ceases the formation of Friedel's salt (Ishida et al. 2008). The usage of fly ash blended binders leads to the formation of additional C–S–H and C–A–H hydrates, which improves the chloride binding capacity of the cement paste. But the carbonation of the chloride bounded cement paste leads to the release of bounded chloride ions into the pore solution. Due to the carbonation, the pH value of the cement paste gets decreases, which makes Friedel's salt to dissolve and results in low chloride binding capacity. In the case of higher carbonation risk, appropriate sealant shall be used for the fly ash based concrete structure to prevent the carbonation process (Kayyali and Qasrawi 1992). In order to improve the chloride binding capacity of the concrete, the usage of upto 50% fly ash is recommended in various studies (Muthulingam and Rao 2016).

The physical and chemical chloride binding capacity of the cement paste containing varying percentage of fly ash can be seen from Fig. 5. It can be seen that, the usage of 20% and 40% fly ash as a binder improved the physical (through C–S–H adsorption) and chemical (through Friedel's salt formation) chloride binding capacity of the cement paste. Whereas at 60% fly ash replacement, the chloride binding capacity of the cement paste gets considerably decreased, which was due to their low Ca/Si ratio resulting from the declination of $Ca(OH)_2$ in the cementitious system (Qiao et al. 2019). From the Fig. 5, it can be noticed that at 0% fly ash replacement, the chloride ions were bounded by the formation of calcium oxychloride. The calcium oxychloride is a phase which results from the reaction between calcium hydroxide ($Ca(OH)_2$) and calcium chloride ($CaCl_2$). The formation of calcium oxychloride is the typical chloride binding mechanism associated

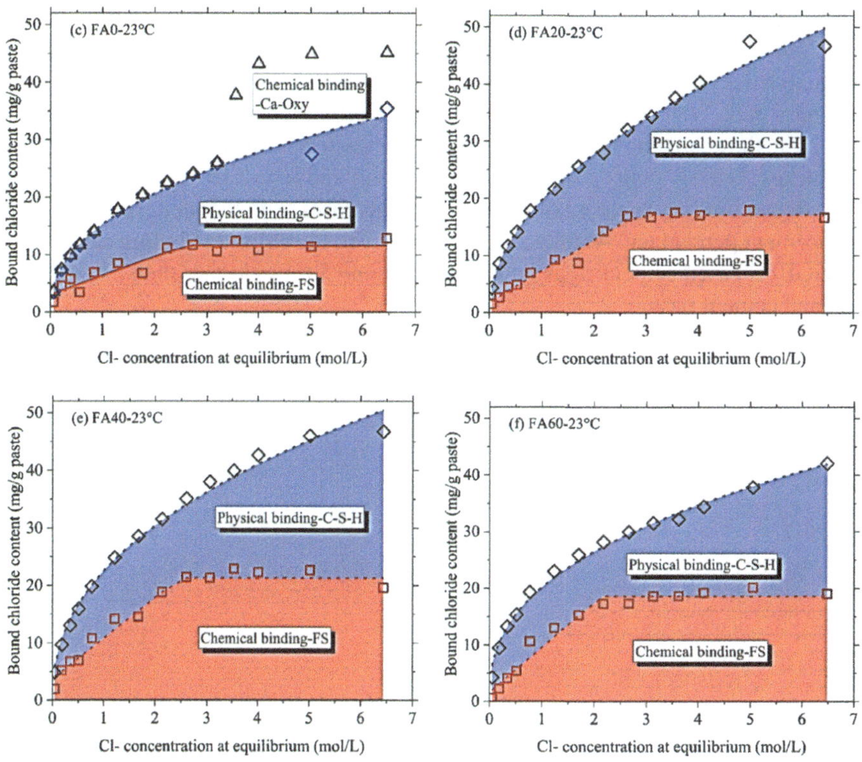

Fig. 5 Physical and chemical chloride binding capacity of cement paste containing various percentage of fly ash at 23 °C temperature (Qiao et al. 2019)

with Type-I portland cement (Qiao et al. 2019). The chloride binding capacity of the high calcium fly ash was found to be higher than that of the low calcium fly ash. It is due to the fact that, the high calcium fly ash possesses early pozzolanic reaction and produce more hydration products which binds the ingressing chloride ions (Sumranwanich and Tangtermsirikul 2004). The usage of 30% fly ash as a binder can improve the chloride resistance of the concrete by 2 to 4 times when compared with the Type-I portland cement concrete (Dhir and Jones 1999).

3.2 Ground Granulated Blast Furnace Slag (GGBFS)

Blast furnace slag (BFS) is generated from the blast furnace, which process the raw iron ore to iron. The blast furnace slags are rapidly cooled with water which leads to the formation of vitrified granulates, due to which they exhibit hydraulic properties, they are ground to the particle size lesser than 100 μm to bring out the cementing properties. So, they can be used in cement production as a partial substitute for the Portland cement clinker (Chandru et al. 2020b). Similar to the other mineral additions, the GGBFS also enhances the mechanical and durability characteristics of the concrete when they are dosed at their optimum replacement level (Chandru et al. 2020c, 2021). Since the average Al_2O_3 content of GGBFS is higher than 20%, they can bind higher chloride ions by the formation more chloroaluminate complexes (Friedel's salt). The study made by Dhir and Jones (1999) reported that the chloride ion binding capacity of the concrete increases with the increase in GGBFS replacement percentage. At 66% replacement level, the chloride binding capacity of the GGBFS based cementitious system was around 5 times higher than the ordinary portland cement paste.

The Fig. 6 shows the chloride content in ordinary portland cement paste and 50% GGBFS admixed cement paste at various depth and time. It can be observed that

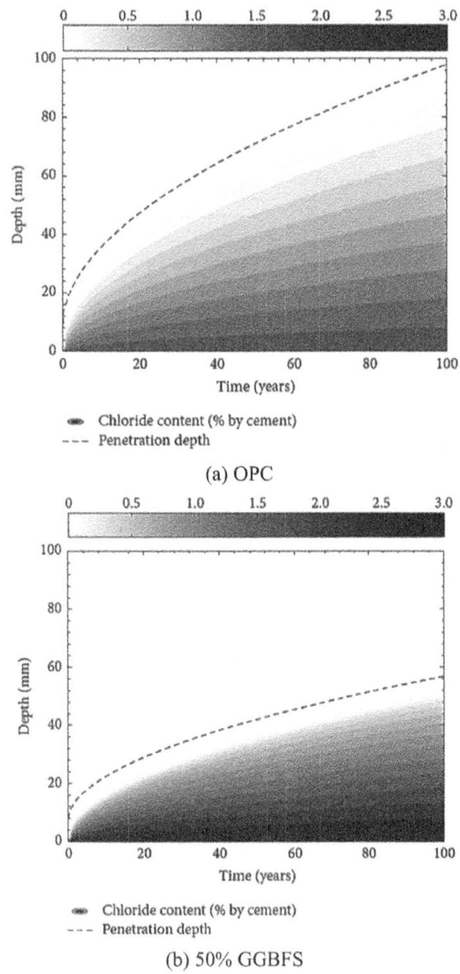

Fig. 6 Chloride content in cement paste at various depth and time (Hong and Ann 2017)

there is a significant decrease in the chloride ion penetration depth with the usage of 50% GGBFS. Since the C_3A content in the GGBFS is higher than that of Type-I portland cement, they possess a better chloride binding capacity (Hong and Ann 2017). The chemical reaction between the aluminate phases and ingressing chloride ion is given in Eqs. (11) and (12) respectively (Kim et al. 2016).

$$C_3A + \text{Chloride ion} \rightarrow Ca_6Al_2O_6 \cdot CaCl_2 \cdot 10H_2O\ \left(\text{Friedel'sSalt}\right) \quad (11)$$

$$C_4AF + \text{Chloride ion} \rightarrow Ca_6Fe_2O_6 \cdot CaCl_2 \cdot 10H_2O\ (\text{Calcium chloroferrite}) \quad (12)$$

The chemical binding of chloride ions follows a substitution mechanism i.e. anion exchange mechanism. The chloride ions in the pore solution gets binds with the AFm phases, which results in the release of equal amount of OH- ions into the pore solution from the interlayers of AFm hydrates. Adsorption of chloride ion by the AFm phases from the NaCl in the pore solution leads to the increase in Na+ ion concentration. In order to balance the ionic charge neutrality of the pore solution, the excess Na+ ions existing in the pore solution enters into the C–S–H phase (Suryavanshi et al. 1996).

When compared with the physical binding (C–S–H adsorption), the chemical binding (Friedel's salt formation) capability to bind chloride ions was found to be higher in the GGBFS admixed cement paste. This is due to the presence of around 20% of Al_2O_3 and significant amount of C_3A content in the GGBFS. But the alkalinity of the pore solution and presence of sulfate ions cause a negative impact on the chloride binding capacity of the GGBFS admixed cement paste (Luo et al. 2003). The chloride binding capacity of the GGBFS admixed cement paste is higher than the chloride binding capacity of the fly ash and limestone admixed cement paste (Kopecskó and Balázs 2017).

3.3 *Silica Fume*

Silica fume is a by-product from ferrosilicon alloy industries. It consists of around 90% of SiO_2 as their chief chemical constituent. The concrete produced with 5–15% of silica fume as a binder showed a better mechanical and durability properties. In the case chloride binding, the performance of the silica fume was reported to be inferior to the GGBFS, silica fume, and metakaolin. The reason behind the weak chloride binding capacity of the silica fume was due to the unavailability of Al_2O_3. Due to the unavailability of Al_2O_3, the silica fume cannot form chloroaluminate complexes (Friedel's salt) with the ingressing chloride ions (Guo et al. 2019). When silica fume is used along with cement, it alters the pH of the pore solution. pH of the pore solution decreases with the increase in silica fume content. Due to the low pH of pore solution, the Friedel's slat become soluble and release back the bounded chloride ions (Page and Vennesland 1993).

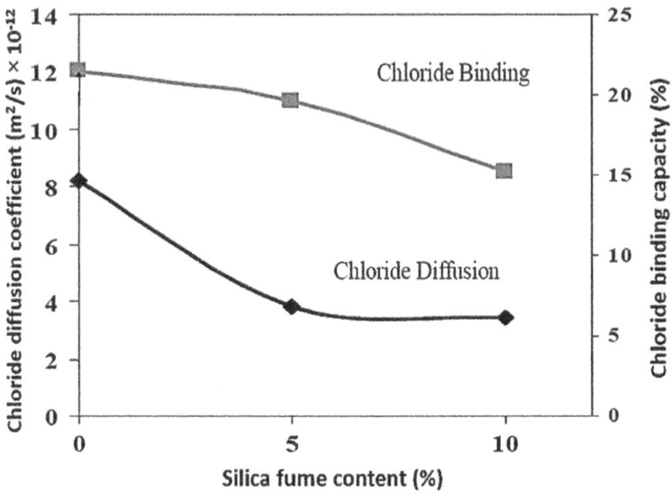

Fig. 7 Chloride diffusion coefficient and chloride binding capacity of silica fume concrete (Dousti et al. 2009)

The usage of silica fume as a partial binder causes the following changes in cement paste, (i) increases the C–S–H content; (ii) decreases the Ca/Si ratio; (iii) decreases the pH value; (iv) decreases the C_3A content. In the above-mentioned changes, the increase in C–S–H content contributes to the chloride binding capacity of the cement paste. Whereas the decrease in Ca/Si ratio, pH value and C_3A content declines the chloride binding capacity of the cement paste. From the Fig. 7, it can be observed that the chloride diffusion coefficient of the concrete decreases with the increase in silica fume content, which was due to the pore refinement effect of silica fume and densification of the cementitious matrix by the secondary C–S–H gel. On the other hand, the chloride binding capacity of the concrete decreases with the increase in silica fume content which was due to the decrease in Ca/Si ratio, pH value and C_3A content in the cement paste (Dousti et al. 2009).

3.4 Metakaolin

Metakaolin is obtained through calcinating the clay-based mineral called kaolinite. The specific surface area of the metakaolin is usually lower than cement particles, but not higher than the silica fume. The metakaolin has around 50–55% of SiO_2 and 35–45% of Al_2O_3 as their chemical constituent which enhances the concrete strength and durability. The presence of SiO_2 promotes the formation of secondary C–S–H gel and the presence of Al_2O_3 improves the chloride binding capacity of the concrete.

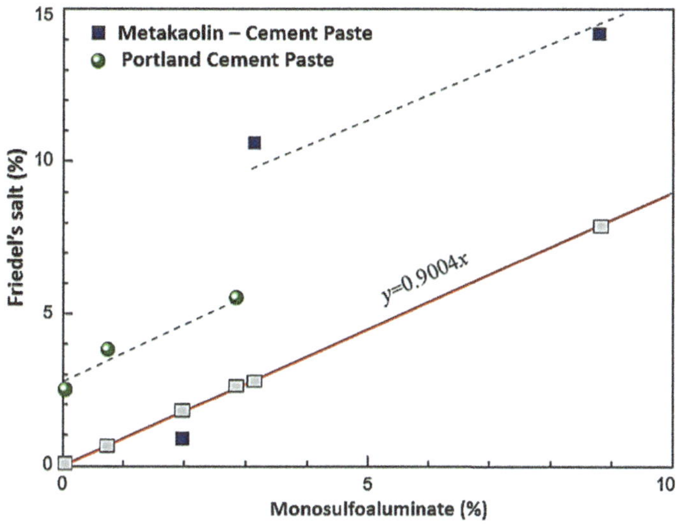

Fig. 8 Formation of Friedel's salt and Monosulfoaluminate in the metakaolin blended cement paste (Gbozee et al. 2018)

Due to the high Al_2O_3 content in the metakaolin, at the optimum dosage, more amount of monosulfoaluminate gets formed. During the chloride ion ingression, the monosulfoaluminate gets converted into Friedel's salt. From the Fig. 8, it can be observed that the increase in monosulfoaluminate content increases the Friedel's salt formation in the cement paste, which in terms enhances the chloride binding capacity of the concrete. It can also be noticed that, due to the low Al_2O_3 content in the portland cement lesser amount of monosulfoaluminate and Friedel's salt gets formed which results in the low chloride binding capacity of the portland cement paste. When the metakaolin was replaced at higher percentage, more ettringite and strätlingite was formed instead of monosulfoaluminate. The ettringite and strätlingite does not much promotes the formation of Friedel's salt (Gbozee et al. 2018).

The influence of metakaolin in the chloride binding capacity of the concrete can be seen from Fig. 9. It can be observed that, increasing the metakaolin content decreased the chloride concentration of the concrete at various depths. But the presence of sulphate decreases the chloride binding capacity of the concrete, sulphate ions reacts with the monosulfoaluminate and forms ettringite which does not possess a good chloride binding capacity (Wang et al. 2014).

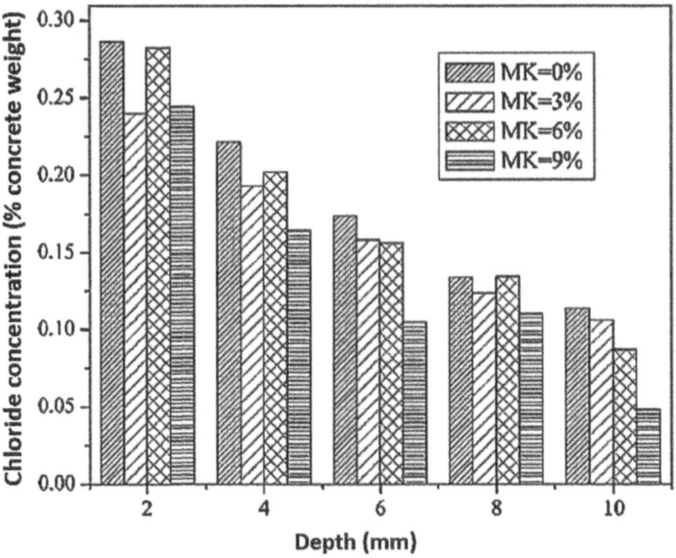

Fig. 9 Chloride concentration at various depths of metakaolin blended concrete (Wang et al. 2014)

4 Conclusions

From the results reported in the various studies, the following conclusions are provided,

- After ingressing into the cementitious system, the chloride ions will be available in three forms, such as chemically bound chloride (Friedel's salt and Kuzel's salt), physically adsorbed chloride (adsorbed within C–S–H layers), and free chloride ions (in pore solution).
- In order to prevent the chloride induced rebar corrosion in the reinforced concrete structures the following three remedial measures can be taken. (i) increasing the AFm phases in the cementitious system to chemically bind the ingressing chloride ions to form chloroaluminate complexes i.e., Friedel's salt and Kuzel's salt; (ii) promoting more C–S–H gel formation to physically adsorb the ingressing chloride ions; (iii) refining the pore structure of the cementitious matrix to make them more impermeable to the gases and aggressive liquids.
- The chloride binding capacity of the concrete depends on the Ca/Si ratio, pH value and C_3A content in the cement paste. In addition to these parameters, the presence of sulphate ions has a considerable influence in the formation of Friedel's salt. In the presence of sulphate ions, more ettringite forms rather than the monosulfoaluminate which affects the chloride binding capacity of the concrete.
- Due to the high Al_2O_3 content in the fly ash, GGBFS and metakaolin, they form more Friedel's salt which results in the better chloride binding capacity. Whereas in silica fume, the unavailability of Al_2O_3 hinders the formation of chloroaluminate

complexes (Friedel's salt) with the ingressing chloride ions and results in poor chloride binding capacity.

References

Chandru, P., Karthikeyan, J., Sahu, A. K., Sharma, K., & Natarajan, C. (2020a). Some durability characteristics of ternary blended SCC containing crushed stone and induction furnace slag as coarse aggregate. *Construction and Building Materials*, 121483. https://doi.org/10.1016/j.conbuildmat.2020.121483.

Chandru, P., Karthikeyan, J., & Natarajan, C. (2020b). Steel slag—a strong and sustainable substitute for conventional concreting materials. In J. Delgado (Eds.) *Sustainable Materials in Building Construction*. Building Pathology and Rehabilitation, 11. Springer, Cham. https://doi.org/10.1007/978-3-030-46800-2_2.

Chandru, P., Karthikeyan, J., Sahu, A. K., Sharma, K., & Natarajan, C. (2020c). Performance evaluation between ternary blended SCC mixes containing induction furnace slag and crushed stone as coarse aggregate. *Construction and Building Materials*, 120953. https://doi.org/10.1016/j.conbuildmat.2020.120953.

Chandru, P., Karthikeyan, J., & Natarajan, C. (2021). Correlations between the Hardened properties of combination type SCC containing UFGGBFS. *Advances in Civil Engineering Materials, 10*(1), 1–22. https://doi.org/10.1520/ACEM20190233.

Cheewaket, T., Jaturapitakkul, C., & Chalee, W. (2010). Long term performance of chloride binding capacity in fly ash concrete in a marine environment. *Construction and Building Materials, 24*(8), 1352–1357. https://doi.org/10.1016/j.conbuildmat.2009.12.039.

Currie, J. A. (1960). Gaseous diffusion in porous media. Part 2.-Dry granular materials. *British Journal of Applied Physics, 11*(8), 318.

Damrongwiriyanupap, N., Li, L., & Xi, Y. (2011). Coupled diffusion of chloride and other ions in saturated concrete. *Frontiers of Architecture and Civil Engineering in China, 5*(3), 267. https://doi.org/10.1007/s11709-011-0112-z.

Dhir, R. K., & Jones, M. R. (1999). Development of chloride-resisting concrete using fly ash. fuel, 78(2), 137–142. https://doi.org/10.1016/S0016-2361(98)00149-5.

Dousti, A., Shekarchi, M., & Tadayon, M. (2009). Effect of silica fume on chloride binding in concrete. In *4th International Conference on Construction Materials: Performance, Innovations and Structural Implications* (pp. 715–720).

Dyer, T. (2014). *Concrete durability*. Crc Press.

Esquinas, A. R., Álvarez, J. I., Jiménez, J. R., Fernández, J. M., & de Brito, J. (2018). Durability of self-compacting concrete made with recovery filler from hot-mix asphalt plants. *Construction and Building Materials, 161,* 407–419. https://doi.org/10.1016/j.conbuildmat.2017.11.142.

Fagerlund G. (1982). *On the capillarity of concrete*. Nord. Concr. Res. 1 6.1–6.20.

Gbozee, M., Zheng, K., He, F., & Zeng, X. (2018). The influence of aluminum from metakaolin on chemical binding of chloride ions in hydrated cement pastes. *Applied Clay Science, 158,* 186–194. https://doi.org/10.1016/j.clay.2018.03.038.

Guo, Y., Zhang, T., Tian, W., Wei, J., & Yu, Q. (2019). Physically and chemically bound chlorides in hydrated cement pastes: A comparison study of the effects of silica fume and metakaolin. *Journal of Materials Science, 54*(3), 2152–2169. https://doi.org/10.1007/s10853-018-2953-5.

Hong, S. I., & Ann, K. Y. (2017). Estimation of corrosion-free life for concrete containing ground granulated blast-furnace slag under a chloride-bearing environment. *Advances in Materials Science and Engineering*. https://doi.org/10.1155/2017/3186371.

Ishida, T., Miyahara, S., & Maruya, T. (2008). Chloride binding capacity of mortars made with various Portland cements and mineral admixtures. *Journal of Advanced Concrete Technology, 6*(2), 287–301.

Kayyali, O. A., & Qasrawi, M. S. (1992). Chloride binding capacity in cement-fly-ash pastes. *Journal of Materials in Civil Engineering, 4*(1), 16–26.

Kim, M. J., Kim, K. B., & Ann, K. Y. (2016). The influence of C3A content in cement on the chloride transport. *Advances in Materials Science and Engineering.* https://doi.org/10.1155/2016/5962821.

Kopecskó, K., & Balázs, G. L. (2017). Concrete with improved chloride binding and chloride resistivity by blended cements. *Advances in Materials Science and Engineering.* https://doi.org/10.1155/2017/7940247.

Li, C., Ma, B., Tan, H., Zhang, T., Liu, X., & Chen, P. (2020). Effect of triisopropanolamine on chloride binding of cement paste with ground-granulated blast furnace slag. *Construction and Building Materials, 256,* 119494. https://doi.org/10.1016/j.conbuildmat.2020.119494.

Liu, J., Liu, J., Huang, Z., Zhu, J., Liu, W., & Zhang, W. (2020a). Effect of Fly Ash as Cement Replacement on Chloride Diffusion, Chloride Binding Capacity, and Micro-Properties of Concrete in a Water Soaking Environment. *Applied Sciences, 10*(18), 6271. https://doi.org/10.3390/app10186271.

Liu, X., Ma, B., Tan, H., Zhang, T., Mei, J., Qi, H., Wang, J. (2020). Effects of colloidal nano-SiO_2 on the immobilization of chloride ions in cement-fly ash system. *Cement and Concrete Composites*, 103596. https://doi.org/10.1016/j.cemconcomp.2020.103596.

Luo, R., Cai, Y., Wang, C., & Huang, X. (2003). Study of chloride binding and diffusion in GGBS concrete. *Cement and Concrete Research, 33*(1), 1–7. https://doi.org/10.1016/S0008-8846(02)00712-3.

Ma, B., Zhang, T., Tan, H., Liu, X., Mei, J., Jiang, W., Qi, H., & Gu, B. (2018). Effect of TIPA on chloride immobilization in cement-fly ash paste. *Advances in Materials Science and Engineering.* https://doi.org/10.1155/2018/4179421.

Meng, B. (1994). Calculation of moisture transport coefficients on the basis of relevant pore structure parameters. *Materials and Structures, 27*(3), 125–134. https://doi.org/10.1007/BF02473025.

Mesbah, A., Cau-dit-Coumes, C., Renaudin, G., Frizon, F., & Leroux, F. (2012). Uptake of chloride and carbonate ions by calcium monosulfoaluminate hydrate. *Cement and Concrete Research, 42*(8), 1157–1165. https://doi.org/10.1016/j.cemconres.2012.05.012.

Muthulingam, S., & Rao, B. N. (2016). Chloride binding and time-dependent surface chloride content models for fly ash concrete. *Frontiers of Structural and Civil Engineering, 10*(1), 112–120.

Page, C. L., & Vennesland, Ø. (1983). Pore solution composition and chloride binding capacity of silica-fume cement pastes. *Matériaux et construction, 16*(1), 19–25. https://doi.org/10.1007/BF02474863.

Plang-ngern, S., & Tangtermsirikul, S. (1999). Chloride binding capacity in fly ash concrete. *Engineering Journal of Research and Development, 10*(2), 1–8.

Qiao, C., Suraneni, P., Ying, T. N. W., Choudhary, A., & Weiss, J. (2019). Chloride binding of cement pastes with fly ash exposed to CaCl2 solutions at 5 and 23 °C. *Cement & Concrete Composites, 97,* 43–53. https://doi.org/10.1016/j.cemconcomp.2018.12.011.

Sumranwanich, T., & Tangtermsirikul, S. (2004). A model for predicting time-dependent chloride binding capacity of cement-fly ash cementitious system. *Materials and Structures, 37*(6), 387. https://doi.org/10.1007/BF02479635.

Suryavanshi, A. K., Scantlebury, J. D., & Lyon, S. B. (1996). Mechanism of Friedel's salt formation in cements rich in tri-calcium aluminate. *Cement and Concrete Research, 26*(5), 717–727. https://doi.org/10.1016/S0008-8846(96)85009-5.

Tang, L., Nilsson, L. O., & Basheer, P. M. (2011). *Resistance of concrete to chloride ingress: Testing and modelling.* CRC Press.

Van Brakel, J., & Heertjes, P. M. (1974). Analysis of diffusion in macroporous media in terms of a porosity, a tortuosity and a constrictivity factor. *International Journal of Heat and Mass Transfer, 17*(9), 1093–1103. https://doi.org/10.1016/0017-9310(74)90190-2.

Wang, G. M., Kong, Y., Shui, Z. H., Li, Q., & Han, J. L. (2014). Experimental investigation on chloride diffusion and binding in concrete containing metakaolin. *Corrosion Engineering, Science and Technology, 49*(4), 282–286.

Yang, Z., Sui, S., Wang, L., Feng, T., Gao, Y., Mu, S., et al. (2020). Improving the chloride binding capacity of cement paste by adding nano-Al_2O_3: The cases of blended cement pastes. *Construction and Building Materials, 232,* 117219. https://doi.org/10.1016/j.conbuildmat.2019.117219.

Comparative Analysis of Numerical Simulations and Experimental Tests for Acoustic Performance in Horizontal Partitions of Buildings

E. C. L. Rezende, A. J. Costa e Silva, A. C. Azevedo, and J. M. P. Q. Delgado

Abstract Acoustic forecasting models from computational simulation (SC) has been the tool used to evaluate the acoustic behavior of the building, even before the work is done, thus enabling the reduction of post-adaptation costs. This research aims to evaluate the reliability of computational simulation by analyzing data from simulations and field measurements (MC) of 14 case studies, in relation to horizontal partition systems (SVH). For the 228 SVH data, of which 116 were for SVH-RA and 102 for SVH-RI, the comparative study showed 38% acceptability between SVH-RA measurement and simulation; and 46% for SVH-RI. This acceptability refers to the values within the ±2 dB range of the aforementioned uncertainty. The data also showed variation between SC and MC, sometimes presenting larger SC, sometimes larger MC. The predicted hypothesis was that SC would always be higher due to the ideal conditions of the simulation, which do not occur in the field. However, where situations of MC greater than SC occurred, it was deduced that deciding on the simulation means choosing to work in favor of safety, since the minimum level of the standard is projected and better results are obtained in the field. It was also verified, through variation of the input data, that the fidelity with the technical specification of the designed systems reproduces reliable values, and the opposite, results in doubtful and discrepant data. In general, the results of the comparative analysis point to the use of computer simulation as positive, since part of the percentage that is not acceptable in this study represents doubtful measurements, and therefore, the

E. C. L. Rezende · A. J. Costa e Silva
Civil Engineering Department, Universidade Católica de Pernambuco, Recife, Brazil
e-mail: emanuelalobo@hotmail.com

A. J. Costa e Silva
e-mail: angelo.silva@unicap.br

A. C. Azevedo · J. M. P. Q. Delgado (✉)
CONSTRUCT-LFC, Departamento de Engenharia Civil, Universidade do Porto, Rua Dr. Roberto Frias, s/n, 4200-465 Porto, Portugal
e-mail: jdelgado@fe.up.pt

A. C. Azevedo
e-mail: antonio.costaazevedo@fe.up.pt

J. M. P. Q. Delgado (ed.), *Building Pathologies and Acoustic Performance*,
Building Pathology and Rehabilitation 18, https://doi.org/10.1007/978-3-030-71233-4_5

simulation is an effective predictive tool and useful in the search for acoustic quality in residential buildings.

Keywords Acoustics · Performance · Acoustic performance · Computer simulation · Acoustic quality

1 Introduction

In urban areas and large cities, numerous pollution problems have negative impacts on the environment and human health. Noise pollution is one of the problems that are potentially being treated by society due to the impact caused both in terms of comfort and the health hazards they cause. To promote better use and occupation of housing buildings, the acoustic quality of the environments has been sought through modifications to consolidated building systems and improvements in building technology. In this context, the NBR 15575 (2013), named as the performance standard, emerges, linking the performance of housing buildings to the aspects of comfort, safety and durability. The standard presents criteria to be met and the manner in which such parameters should be measured. Since acoustic performance is one of the requirements of the standard, the aim is to attenuate sound between environments, whose objective has involved not only builders, but also designers, material suppliers and users.

1.1 Acoustic Comfort

The term "comfortable" has its origins in the Latin word confortare, which emerged to designate actions of "strengthening" and "comfort" […] notion of "support" that has been expanded over the years, going on to mean something "tolerable" or "sufficient", acquiring, only from the 18th century, the sense of "physical well-being" and "pleasure" (Mahmood 2017).

Although the idea of comfort has several perspectives, the concept covers physical-constructive and psychological aspects, and comfort can be perceived, felt, experienced and even measured, classified and standardized, according to Freitas (2005). When talking about acoustic comfort, it is soon associated with the control of disturbance of quiet due to noise. Noise that is associated with unpleasant sounds, i.e., noises.

But what in fact would be acoustic comfort? The concept of acoustic comfort converges to the sensory domain of sound perception, as it suggests a sense of emotional tranquility, privacy at home and in personal activities, satisfaction in work and leisure activities. Often, the feeling of comfort is not clearly noticeable a priori, but the reverse, that is, the discomfort caused by its lack, of ready is manifest.

According to Neto (2009), acoustic comfort can be a subjective concept or an objective requirement, and in one way or another, it has been increasingly demanded by owners or users of buildings. The formality of the concept, however subjective, may not be understood by the population, however the absence of acoustic comfort is increasingly noticeable and the objective requirement to live or work in acoustically comfortable environments is becoming more frequent. Neto (2009) in his study argues that privacy is directly related to acoustic comfort, so that in her work the author investigates what level of comfort would be ideal to minimize such disorders to family life, her rest and relationships between neighbors.

The definition of comfort as a subjective concept has already been studied by several researchers. Rasmussen (2004) explains that acoustic comfort can be understood by the absence of unwanted sound and by activities that do not cause people to be intheir. It also highlights that, just as it is unpleasant to show exposure to noise produced by neighbors, knowing that their activities can be heard characterizes discomfort and acoustic discomfort. Neto (2009) highlights that the concept covers several factors and relates it to peace, privacy, well-being and quality of life, a definition similar to that of other authors, such as Freitas (2005) who says that comfort, in general, is related to psychological issues of identification and satisfaction with the place, as well as to physical temperature conditions, humidity, ventilation, lighting and acoustics.

Berrio and Castro (2012) clarify that comfort is a multidimensional dynamic function of several independent variables, affected by subjective perceptions, views and customs. Thus, acoustic comfort is not a purely objective concept, since the particularity of discomfort is a variable among people, as illustrated in Fig. 1.

It is noted that acoustic comfort is only remembered when the discomforts begin to emerge, an example given in the case of the acquisition of the residences studied by Neto et al. (2010), where currently more and more residential developments are sought with vast leisure options, with well divided and spacious internal spaces, such as a true private leisure club; and the discussion about acoustic comfort only arises when, after the occupation, the residents begin to listen to the conversations of the

Fig. 1 Peculiarity of the question of acoustic nuisance-contrast of perceptions

neighbors, the steps on upper floors, that is, when the noise begins to hinder the coveted quiet. Jensen et al. (2019) show that discomfort with neighborhood noise is associated with several factors of mental and physical health impairment.

Very rarely the issue of acoustic comfort was being considered when designing an enterprise, and therefore, the solutions in the post-work are always more expensive and less efficient. However, the lack of priority in the issue of acoustics in projects is not the exclusive fault of professionals, since rules and guidelines of acoustic comfort for indoor environments and the concern with good sounding are relatively recent themes in Brazilian standards for housing buildings, when compared to the history of civil construction, as Neto (2009) points out.

Research shows that, in addition to the construction method of buildings, the materials employed, the decorative objects and human occupation, that is, all the components of the building contribute to estimate the overall comfort of the building, as Silva (2012) explains, which complements highlighting that each type of material and construction system has a peculiar acoustic characteristic, reinforcing that to evaluate the good acoustic insulation of the building systems and whether it has good ability to absorb sound, the ideal is the laboratory measurement of the level of sound insulation (IA), whose some examples are presented in Table 1.

By the values of this table, it can be observed that the sound insulation is both higher as the density of the material of the construction system adopted, which proves the relationship of acoustic insulation with the law of the masses, described by Souza et al. (2013). However, Silva (2012) explains that the insulation index does not fold in the same relationship that doubles the density of a material, due to the logarithmic function that determines such index, concluding that acoustic attenuation increases by 6 dB when the surface density is doubled, as presented in Eq. (1),

$$IA = 20log(\delta S) \tag{1}$$

where IA is the sound insulation or attenuation index; and δS is the surface density.

Table 1 Sound insulation index (IA) of building materials/systems

Materials or systems	IA (dB) 500 Hz
Concrete wall 30 cm, with a big aggregate, towed	50
Concrete wall 15 cm, w/aggregate kid	40
Massive brick wall or stone 45 cm	55
Solid brick wall 22.9 cm, with 13 mm plaster	50
Solid brick wall 11.2 cm, with 13 mm plaster	45
Cell concrete wall siporex 15 cm, density 450 kg/m^3	42
Siporex cell concrete wall 7.5 cm, density 450 kg/m^3	35

In this sense, seeking acoustic comfort presupposes adopting insulation solutions and in this sense, some standards were created to guide the evaluation of acoustic intensity.

Within the scope of the legal basis that protects in Brazil regarding acoustic comfort, the standards NBR 10151 (2019) and NBR 10152 (2017) were created in 1987 as a reference for noise evaluation, aiming at the comfort of the population. However, they are not supervisory standards, but reference standards, in order to give conditions to assess the adequacy of an environment according to its use regarding the issue of acceptable noise.

The perception of neighborhood noise is an important indication of this acoustic discomfort. Characterized by sound transmission between contiguous environments, neighborhood noise contributes to the potentiation of acoustic incomocity inside homes. The internal transmission of these noises, either through the walls and slabs of the environments or by the shock of an object in a rigid structure (previously cited as the impact sounds), presupposes the need for sound isolation in order to mitigate such discomfort. Isolation that is treated in the performance standard NBR 15575 (2013), as a mechanism that contributes to the improvement of the acoustic comfort of the environment and the search for well-being and quality of life of people, after all comfort and performance are concepts that complement each other.

According to reference values for comfort, Table 3 of the NBR10152 (2017), in a residence, determines that the overall sound pressure level of a dormitory must be less than or equal to the weighted average index of 35 dB (R_{LAeq}), or if the sound source within the environment the maximum level can reach 40B (R_{LASmax}), being considered as the representative NC level of the dormitory environment the value of 30. Measured or projected values above these references are considered to be unacceptable, i.e. not comfortable.

As for performance, NBR 15575 (2013) does not present representative values of acoustic comfort of the environment, but values that guide sound insulation between walls and floors, that is, acoustic attenuation values. For the same example used in a dormitory environment, by NBR 15575 (2013) the minimum insulation required between walls of autonomous units, that is, between two apartments in which at least one of the environments is a dormitory, should be 45 dB, exactly in order to minimize the noise transmitted and seek acoustic comfort, because as cited performance and comfort complement each other.

Several works published by researchers around the world, cited by Neto (2009) in his research, point out several factors causing the discomfort of noises, the relationship of audibility with the situation of discomfort, in addition to mentioning the noises of neighborhood, coming from impact of doors in knock, impact of objects on the floor, walking and children playing, radio and TV sounds at high power, barking of dogs, in addition to conversation between people, as the reasons that cause the most discomfort in residential buildings. Finally, they reveal the tendency of neighborhood noise, in some cases, to overcome traffic noise, according to one of the case studies cited.

Thus, it is perceived that it is evolving in the search for the best quality of life and comfort in buildings, both by the standardizing bodies and by the professionals involved in research and work on the subject, because the issue of noise, whether the noises from the outside of the building or those coming from internal activities, continuously contribute to the acoustic discomfort of the environments, even impacting on the physical and mental health of individuals.

In conclusion, when talking about comfort, we talk about human need, because the definition converges to the state of well-being of the individual, necessary to guarantee the quality of life, while the concept of performance refers to the behavior in use of a given element, system, etc.

1.1.1 Normative References Regarding Comfort—NBR 10151 (2019) and NBR 10152 (2017)

The NBR 10151 (2019)—Acoustics: Evaluation of noise in inhabited areas, initially published in 1987, revised in 2000 and, recently, in 2019, comes to deal with the levels of evaluation for external and internal environments, determining criteria and methods of measurement and establishing ranges of acceptable noise values depending on the day or night period of the day.

The NBR 10151 (2019) standard sets required conditions for assessing the acceptability of noise in the community, explaining a method for noise measurement (Arndt et al. 2010). The levels of evaluation criteria (NCA) are the maximum sound pressure levels admitted during the day and night in the communities, varying according to the type of area and the period of noise incidence. The standard brings to each type of external environment a different NCA for the day and night period, explaining that, for indoor environments, these NCA values are corrected by -10 dB(A) and -15 dB(A) for open or closed windows, respectively, that is, according to the conditions of use of the environment.

With the new update of 2019, the standard brought more details for measurement and evaluation of sound pressure levels in inhabited areas for general application. Table 2 presents the NCA for external environments, expressed in dB(A) which is the weighted unit of measure defined by the standard.

The NBR 10152 (2017)—Acoustics: Noise levels for acoustic comfort, from 1987, which originated the NB-95/1986 project, sets the noise levels compatible with acoustic comfort in various environments, establishing acceptable values for

Table 2 Assessment criterion level (NCA) for external environments, in dB (A)

Types of areas	Daytime	Night
Site and farm areas	40	35
Urban or hospitals or schools strictly residential area	50	45
Mixed area, predominantly residential	55	50
Mixed area, with commercial and administrative vocation	60	55
Mixed area, with recreational vocation	65	55
Predominantly industrial area	70	60

such environments, from which there is talk of discomfort, strongly conditioning the productivity and even the health of people. However, this standard does not deal with the risks of damage to health due to noise, a subject studied in specific standards, but rather the mean, maximum and representative reference values of sound pressure admitted for each type of environment, respectively R_{LAeq}, R_{LASmax} and NC.

According to NBR 10152, published in 1987, the acceptable deck limit depends on each environment, since the requirements of a hospital cannot be the same as a meeting room or a sports venue. However, as mentioned, the standard is not a supervisor of a specific sound, but rather a limiting of the total sound for a given environment.

Updated after 30 years of the initial release, NBR 10152 (2017) has been called NBR 10152—Acoustics: sound pressure levels in internal environments to buildings, expanding the approach of the previous version and brought, in 2017, more complete specifications, with new reference values for internal environments to buildings, and procedures for measuring and evaluating sound pressure levels with new table of criteria of internal noise and incorporation of previously unmentioned environments, but with the same focus on ensuring that sound activities remain within a healthy standard of intensity and volume, that is, to allow an international standard to environments where the issue of acoustic comfort is fundamental.

In addition to the representative sound pressure level, called the comfort level (NC), the standard presents measurement orientation in two different methods, one simplified and one detailed, and R_{LAeq} reference values correlated with the standardized weighted level difference values of the NBR 15575 (2013) performance standard.

In the previous version of NBR 10152, published in 1987, the representative comfort level of an internal environment of a building (NC) was graphically determined from the noise evaluation curves, obtained through the relationship between the equivalent sound pressure level dB and the central frequencies of the Octave Hz bands, as explained by Silva (2012).

In the updated version, 2017, NBR 10152 brought these values of the nc noise curve tabled, for each band of 1/1 octave, explaining that the determination of the representative level of the LNC environment is given by the highest NC value identified between the comparison of all representative levels of equivalent sound pressure L_{eq} (HZ) to each band, thus facilitating the reading of the NC representative value of the environment.

The new version 2017 brought in a complementary way another table with interpolation of these curves every 1 dB, in order to increase the use of NC levels.

Table 3 is the reproduction of the table of NBR 10152 (2017) which contains the mean, maximum and representative reference values of sound pressure admitted for each type of environment reviewed, used for the elaboration of studies and acoustic projects with a view to acoustic comfort. It is noteworthy that NBR 10152 (2017) is not limited to residential buildings such as NBR 15575 (2013), because the comfort standard covers other types of real estate such as commercial, health, educational, leisure, commercial and other types.

Table 3 Reference values for internal environments of a building, NBR 10152 (2017)

Purposes of use	Reference values		
	RL_{Aeq}(dB)	RL_{ASmax}(dB)	RL_{NC}
Airports, bus and train stations	45	50	40
Check-in areas, ticket offices	50	55	45
Departure rooms and circulations			
Shopping centers	50	55	45
Circulations	45	50	40
Stores	50	55	45
Food stalls	55	60	50
Garages			
Clinics and Hospitals	35	40	30
Nurseries	35	40	30
Surgical centres	35	40	30
Offices	40	45	35
Wards	45	50	40
Laboratories	40	45	35
Collective rooms	35	40	30
Individual rooms	45	50	40
Waiting rooms			
Cultural and Leisure	40	45	35
Party halls	45	50	40
Restaurants	35	40	30
Cinemas	30	35	25
Concert halls	30	35	25
Theaters	40	45	35
Small religious temples ($\leq$600 m^3)	35	40	30
Large religious temples (>600 m^3)	40	45	35
Libraries	40	45	35
Museums (exhibitions)	25	30	20
Recording studios			
Educational	50	55	45
Nursery	40	45	35
Circulations	35	40	30
Classrooms	35	40	30
Music rooms			
Offices	50	55	45
Call centers	50	55	45
Circulations	40	45	35
Private offices (management, board, etc.)	45	50	40
Collective offices (open plan)	45	50	40
Receptions	45	50	40
Waiting rooms	35	40	30
Meeting rooms	40	45	35
Videoconference rooms			
Sports	45	50	40
Sports gyms and gyms			

(continued)

Table 3 (continued)

Purposes of use	Reference values		
	RL_{Aeq}(dB)	RL_{ASmax}(dB)	RL_{NC}
Hotels			
Single rooms or suites	40	45	35
Convention halls	40	45	35
Service areas	50	55	45
Circulations	45	50	40
Residences			
Bedrooms	35	40	30
Living rooms	40	45	35
Home theaters	40	45	35
Others			
Large auditoriums (>600 m^3)	30	35	25
Small auditoriums (≤600 m^3)	35	40	30
Kitchens and Laundries	50	55	45
Courts	40	45	35

1.2 Building Performance

Performance is understood as behavior in use, and has a direct relationship with what is expected of something, with the expected yield according to the function for which it was created. Performance of the buildings goes back to the concept of service life and durability of the buildings, since when designed, it is expected that the enterprise offers satisfactory conditions to the user, as calculated and desired.

In a study that sought to cover the conceptual evolution of the theme performance of buildings in Brazil, Borges (2008) demonstrates that the construction sector seems to understand superficially the importance of the theme, since with the standardization, quite recent, it has been limited to compliance only with minimum levels. However, it demonstrates that the theme in the world has focused on aspects related to sustainability since the 1970s. In this approach, the author points out that the application of the concept of performance is a global trend, appropriate to the treatment of environmental aspects and service to users and that in itself requires an environment favorable to technological innovation, thus representing an opportunity to improve the quality of Brazilian homes and optimize resources.

1.2.1 Concept History—Performance

The word performance is used colloquially by society associating to a desired level of quality and behavior. Group of people, lovers of cars and electronic equipment, sports commentators, team leaders, human resources specialists, business analysts, among others, use in their daily life the word performance. What is common among groups is that the desired performance definition is compared to the result delivered (Borges 2008).

The effective application of the concept began in the period of World War II, when the manufacture of products destined to the war industry had to meet the requirement of structural security (see Borges and Sabbatini 2008).

Regarding the performance of buildings, the concept has been studied since the 1960s, based on questions presented at the second congress of the Council International for Building (CIB), an organization dedicated to the advancement of construction technology. The definition of performance of buildings was associated with their behavior when in use, related throughout their useful life [...] And the building, as a product that presents certain characteristics that enable it to meet specific objectives and functions, is considered as "well-behaved" when it meets the requirements for which it was designed (Borges 2008).

The concept of performance involves human needs that must be met by the building in order for it to fulfill its function throughout its useful life. (Kern et al. 2014). Discussing the evolution of the concept, Borges (2008) cites the Hammurabi code of Babylon between 1955 and 1913 BC, where there is no talk of the technique of how to build, however it is understood that there is a clear regulation that construction cannot collapse, thus characterizing the desired performance in construction at that time.

> Article 229: The builder made a house for a man and his work was not strong, and if the house falls and kill the owner of the house, then the builder must be killed (CODE HAMMURABI).

The theme performance began to be studied by several important entities in the world. In the United States, through the U.S. Department of Housing and Urban Development, in the 1970s, through a program aimed at developing criteria for projects and evaluation of innovative systems for the construction of homes, which resulted in the publication of a document in 1977 with performance criteria.

At the same time, in 1970, CIB created commissions and working groups focusing on issues focused on the environment and construction themes for society. The CIB W60—The Performance Concept in Building committee aimed to establish a conceptual and technological structure on the performance of buildings that could be adopted at the international level (Kern et al. 2014).

From then on, the theme "performance for constructions" began to be valued and studied more deeply, resulting in the concept given by Gibson (1982): *"First and foremost, the performance approach is [...] the practice of thinking and working in terms of ends rather than means.[...] It is concerned with what a building or building product is required to do, and not with prescribing how it is to be constructed."*

To consolidate the concept, from 1972 to 1996, international conferences were held in countries such as Philadelphia, Finland, Portugal, United States, Luxembourg, Israel, sponsored by CIB, ASTM, RILEM and ISO; and since 1996, annual seminars in several countries discussed the practical application of the concept of performance. In the 1990s, European countries such as Denmark, the Netherlands, Ireland and Spain began to adopt performance measures for buildings linked to the ratio of performance in energy consumption. [...] In 2000, the PeBBu Theme

Network (Performance Based Building) was created in the European Union with the purpose of consolidating the work on performance, being percussors of technical standardization directives (Kern et al. 2014).

The concept of performance thus, linked to meeting an expectation generated, especially to human needs, demonstrates a direct relationship with the useful life of buildings, to the extent that it involves issues of durability and sustainability, themes of great repercussion in the construction industry, as an articulator of the longings for habitability and comfort of society. When the buildings have less durability, presenting pathologies and requiring repairs or demolitions, it is said that the performance of the buildings was compromised, thus enhancing the environmental impact generated by civil construction. In addition, these problems and pathologies in housing buildings can result in premature obsolescence of buildings and their surroundings, as advocate Boselli and Dunowicz (2009), which lead to the decline in quality of life, durability of buildings and safety of residents.

Thus, the relationship between performance and sustainability is of paramount importance in the current scenario, especially given that the construction industry is presented as the human activity with the greatest impact on the environment, either by the high consumption of natural resources, by the major changes in the landscape and also by the high generation of waste (Agopyan and John 2011).

1.2.2 Standardization—Performance

The elaboration of international standards ISO 6240 and 6241, in the years 1980 and 1984, respectively, represented the framework of application of the concept of performance in buildings, as it brought general principles for the preparation of performance standards in construction, assisting committees and working groups of standards concerned with the performance of entire buildings, parts of buildings and external spaces, to meet the specific requirements of users and the expectation of society. The objective of the standards, according to Borges (2008), was to serve as sistance to countries that followed ISO in the elaboration of their own performance standards, serving as a guide for the choice of specific requirements for service to users, applied on a case-by-case basis.

ISO 6241 considered aspects of stability, fire safety, safety in use, partition of parts against weather, temperature control, ventilation and noise control, aspects of luminosity of environments, as well as tactile and dynamic properties of components, hygiene, maintenance of buildings and durability and service life; important items that were a reference of the Brazilian Standard NBR 15575 (2013) but that were already addressed 29 years before its publication in 2013.

In Brazil, in the 1960s, public policies to encourage the implementation of housing of social interest resulted in the creation of the National Housing Bank—the BNH, whose initial objective was to guide and control the Housing Financial System, but promoted the construction and acquisition of homeownership, thus encouraging the development of new construction systems.

The industrialization of construction and rationalization, results of economic growth and incentive to investments in the sector, enhanced new technologies and new construction systems, however there were no technical assessment instruments or criteria for analyzing such solutions and their repercussion over the expected useful life of buildings (Borges 2008).

In Brazil, the concept of performance in construction is strongly linked to housing. One of the first presentations of the concept of performance at the national level occurred through the academic work of Professor Teodoro Rosso, in the 1970s, at the Faculty of Architecture of the University of São Paulo (Borges 2008). But the first official publication takes place around the 1980s, the result of research by the Institute of Technological Research of the State of São Paulo (IPT) together with the National Housing Bank (BNH), with the purpose of elaborating criteria aimed at evaluating the quality of construction systems and new products. This publication, called "Minimum Performance Criteria", aims to fill the existing standardization gap of performance criteria evaluation, due to the scarcity of references focused on the specific needs of Brazil, and consequently disastrous results of application of techniques that are poorly consolidated or unknown nationally (Borges 2008; Kern et al. 2014).

At that time, all Brazilian normalization was prescriptive and, for the most part, focused on the specification of constructive solutions without determining minimum acceptance limits that served as a reference to the performance evaluation of such solutions.

The development of studies on the performance of housing buildings came from the need to ratify the effectiveness of construction systems to the detriment of the low quality of the final product (SEBRAE 2016). With the extinction of the National Housing Bank in 1986, its replacement, Caixa Econômica Federal resumed in 1997 the studies in contract with the IPT, in line with other studies that had been done by other reputable entities of the technical environment, such as the Brazilian Institute of Technology and Construction Quality and other. Thus, in 2000 these various references were harmonized by the research "Technical Standards for evaluation of innovative building systems for housing", with the support of the Studies and Projects Financier—Finep, whose objective was to create a method of evaluation of innovative construction systems, that is, the development of a set of Brazilian technical standards for evaluation of housing buildings based on the principles of the performance concept. This was the kick of the creation of the NBR 15575 (2013) performance standard.

Based on a national and international literature review, the research raised questions of difference between classes of buildings and consequently difference in users' requirements according to the function of buildings, issues of evaluation of the building as a whole and in isolation in their systems, which are structural, hydraulic and sanitary, walls and others.

In 2004, already with ready-made basic texts, actors in the production chain of civil construction: builders, developers, designers, university, research institutes, manufacturers of materials and components; including government representatives with the Ministry of Cities, Caixa Econômica and the Brazilian Chamber of Construction

Industry—CBIC, discussed in public plenary at the Polytechnic School of São Paulo what was officially published as the Performance Standard on May 12, 2008 (Borges 2008; Kern et al. 2014). However, the application of the standard was postponed to 2013, after the republication of the 2008 edition, with a view to demystifying the impacts arising from the targets to be achieved with the fulfillment of this norma.

The great challenge in using the performance approach in construction was to translate users' needs into requirements and criteria that could be objectively measured (Borges 2008).

1.2.3 Performance Standard—NBR 15575 (2013)

The first edition of NBR 15575 was made available in 2007 for public consultation for publication and application in 2008, however, due to the various questions and the need for corrections and improvements, the application of the standard was postponed and finally published and applied on July 19, 2013, representing a milestone for society and the Brazilian housing and construction market. With attributions imposed, in an individualized way with regard to the effective action of each intervener, including the user in this scenario of responsibility for full compliance with the requirements, the standard represents an advance within the construction cycle of buildings (ASBEA 2015; Piffer 2013; Sinaenco 2015).

As defined by NBR 15575 (2013), Performance Standard is a set of requirements and criteria established for a housing building and its systems, based on user requirements, regardless of its form or constituent materials. It differs from the prescriptive rules, the requirements and criteria of which are established on the basis of the consecration of the use of a particular product or procedureao longo do tempo.

NBR 15575 (2013) introduced a new paradigm with regard to civil construction standards. Its starting point is the user experience and not the constructive technology used. It defines the expected result from the consumer's perspective and not from the perspective of the employed market (Carraro and Jucá 2014).

Unlike traditional standards, which prescribe characteristics of products on the basis of the consecration of use, taking into account their characteristics; performance standards define necessary properties of the different elements of the construction, regardless of the constituent material, thus meeting the needs of the construction (CBIC 2013).

According to the definition itself, the essence of the Standard is the behavior in use of the elements and systems of the building focusing on meeting the requirements of users and not on the prescription of the way of execution of construction systems. From this perspective, NBR 15575 (2013) was organized taking as reference the conditions of implementation and exhibition of the building and the requirements of users, thus fixing:

- qualitative criteria to be met, which were called "Requirements";
- quantitative quantities to guarantee such qualities, called "Criteria", also establishing minimum, intermediate and higher levels of care;

- and uniform forms of evaluation of these parameters, through laboratory and in-field tests, numerical simulations, prototypes and project analyses.

According to Studies of CBIC, the performance standard establishes levels of safety, comfort and resistance in each of the systems that make up a residential building, such as structural systems, floors, internal and external vertical partitions, roofs and hydrosanitary installations. It also seeks to establish parameters, objectives and quantitative, aiming at the discipline of the relations between the links of the economic chain (traceability), the reduction of uncertainties of subjective criteria (expertise), the instrumentation of the Consumer Protection Code, the stimulus to reduce predatory competition and an instrument of differentiation of companies (CBIC 2013).

As SEBRAE (2016) points out, an NBR (Brazilian Standard) is not law, but by law, the standard is mandatory, so that it has the purpose of ensuring the minimum quality conditions that a dwelling must have. The NBR 15575 (2013) is applied only to buildings of the housing type, with any number of floors, new, that is, from the date of entry into force of the standard. It means, then, that it does not apply to works already completed, pre-existing constructions, works in progress, or to projects filed in the competent bodies until that same date, renovation works or retrofit and provisional buildings.

Therefore, the obligation of service refers to everything designed and executed from July 19, 2013. To meet user requirements, the standard is organized around pillars of safety, habitability and sustainability, establishing, over thirteen disciplines/factors, the requirements and consequently the criteria that the building must achieve in use.

Safety requirements are defined by structural safety, fire safety and safety requirements in use and operation. For habitability requirements, the requirements are watertight; thermal performance; acoustic performance; lumine performance; health, hygiene and air quality; functionality and accessibility, and lastly, tactile and anthropodynamic comfort. Sustainability requirements involve durability, maintainability and environmental impact factors.

Some definitions, not mentioned so far in the present study, deserve to be presented for a better understanding of the terms mentioned in the Standard. According to the CBIC (2013) guide it is necessary to:

- "COMPONENT: Integrated unit of a certain building element, with defined shape and intended to fulfill specific functions (e.g. masonry block, tile, door sheet).
- ELEMENT: Part of a system with specific functions. It is usually composed of a set of components (e.g. masonry partition wall, prefabricated partition panel, covering structure).
- SYSTEM: Most functional of the building. Set of elements and components intended to comply with the macro function that defines it (e.g. foundation, structure, floors, vertical partition, hydrosanitary installations, roofing)."

It is important to highlight that the performance analysis performed through NBR 15575 (2013) is for the building construction systems and not for the elements or components alone, because evaluating a material type or a constructive element alone is not sufficient to evaluate the overall performance of an environment, since the interaction between the components and elements is that allow the definition of the different functions of the building building construction systems. Divided into six parts, NBR 15575 (2013), the Performance Standard addresses general requirements of the work, structural system, flooring system, roofing system, partition systems and hydraulic systems.

The standard itself does not address the standards of constructive and design solutions, but addresses different levels of compliance with quality requirements, depending on the basic needs of safety, health, hygiene and economy. Thus, the standard establishes three levels of performance for the different systems: minimum level, intermediate level and higher level; but requires only the minimum level of performance as mandatory.

This is a controversial point in the norm: the obligation to attend to the minimum level, because the performance is fully related to the durability and useful life of the building and, when relating performance with time. It is possible to observe in a graph performance versus time that the tendency is that the initial performance decays with time, being re-established from the maintenance carried out. For this reason, easing the performance buildings at a minimum level, which is required, will not guarantee that, over time, the estimated performance in the design phase maintains the required requirements conditions, i.e. the required performance of the building over its useful life.

Thus, when talking about the performance of a building, one thinks about the service of the estimated useful life of that enterprise, logically, at least the service to the service life of the project, VUP. As a project life, it is understood as the period of time during which the building can be used under satisfactory conditions of safety, health and hygiene and not ending at the time when the building is disposed of to the user. For this reason, much is said that the performance standard reinforces the issue of the durability of the building over time, as it is sought along with prescriptive norms, promoting, throughout the total life of the building, quality characteristics that can be measured and met. That is, even after being delivered to the user, each subsystem of the building will have a horizon to be fulfilled maintaining its performance, provided that the prescriptions of use, operation and maintenance are observed. This reinforces the discussion of maintainability present in the standard and suggests further discussions regarding the minimum level of quality required for some of the requirements of the standard.

According to Crusius (2015), the main agents responsible for implementing the performance standard in the construction industry are construction companies, developers and architectural designers, but given the complexity of the themes, this implementation has been a challenge to these agents, especially the designers.

The implementation of performance standard policies is a highly complex and challenging process. Kern et al. (2014) reinforce this complexity by comparing the process of implementing performance standards in Spain (Technical Building Code) and Brazil (NBR 15575 2013), seeking to identify strategies used in the European country that enabled and facilitated the use of the rules by professionals, suppliers and construction actors. They concluded that, contrary to what was established in Brazil, where the implementation of the entire standard is required, in Spain the implementation took place gradually, in an interactive environment with broad communication between professionals and the Technical Code team, which allowed adjustments, improvements and exchange of experiences, in addition to the cost exemption in the acquisition of said standard, a major obstacle to the Brazilian case.

In Brazil, due to such an extension of the standard, and the complexity/variability of so many aspects, several entities have published explanatory manuals of the standard in order to assist the compliance and adequacy of projects and works, such as CBIC, ASBEA, CAU, Sinaenco, ProAcústica, etc.

NBR 15575 (2013) will generate a positive impact not only on the housing construction segment, since the requirements defined by it have been observed and valued by users of other types of buildings (Sinaenco 2015).

Thus, researchers point as a tendency to approach the theme of performance and the premises of the norm to other types of buildings. Although it exclusively addresses housing buildings, it is expected that it will serve as a reference for the construction of other types of buildings and stimulate the development of specific standards (Piffer 2013), since the performance over time of a building is a concept tied to the theme durability, and this is not an exclusive concern of housing buildings, but of the entire building construction chain.

Acoustic Performance Restricting the object of this work to the acoustic performance requirement, the theme is addressed in part 3 and 4 of the standard NBR 15575 (2013). Knowing that sound waves can reproduce by different means, and that acoustic phenomena of diffraction, reverberation, resonance and absorption interfere in the acoustic performance of environments, the norm is not related to studying these phenomena, but is restricted to the study of the attenuation of the propagation of these sounds, as published by CBIC (2013).

The performance standard does not establish acoustic comfort criteria, otherwise it presents criteria that allow reasonable performance conditions, translated into air sound insulation values transmitted through external partitions (in this case, through facades), to the air sound transmitted between the walls of neighboring apartments, and also for isolation to noise transmitted by the impact between different units, this item of extreme importance in the issue of between floors and roofs.

Although they do not specifically mention acoustic comfort, the values presented of acoustic performance level should be compatible with the noise level of the site of implementation of the enterprise, observed the levels of evaluation criteria—NCA, determined by the standard NBR 10151 (2019), besides being within the limits of noise levels—NC, compatible with the comfort of the internal environments, specified in NBR 10152 (2017).

The definition of acoustic performance contained in NBR 15575 (2013), is that the housing building must present adequate acoustic insulation of the external partitions, with regard to air noise stemming from the exterior of the housing building, and adequate acoustic insulation between common and private areas.

The systems, their materials, components and elements, must provide soundproofing between different units and between enclosures of the same unit, when intended for night rest, home leisure and intellectual work, ensuring minimal acoustic comfort, in terms of levels of background noise transmitted by the airway and structural, and acoustic privacy, in terms of non-speech intelligibility (Ribas 2013). The requirement of acoustic insulation, qualitative parameter, is divided according to the source of the noise source, that is, noise whose source is outside the building and noise coming from inside the neighboring units, as exemplified in Fig. 2.

For acoustic insulation of external partitions, the requirement is to provide minimum performance conditions with respect to normalized sources of external air noise. For acoustic insulation between environments, both between floors and between walls, the standard proposes that there are acoustic insulation conditions between the common areas and environments of housing units and between different housing units, not referring to the environments of the same housing unit. And to meet the requirements of insulation to impact noise, the standard has as a requirement the minimum acoustic performance conditions inside the building, with respect to standardized sources of impact noise.

In Brazil, until approximately 2013, there were no normative instruments that defined mandatory minimum values for sound isolation between housing units (Bezerra et al. 2018). These indexes, which characterize the acoustic performance criteria of a building, should be met already in the project phase of the enterprise, which according to Patrício (2005, 2010) is what occurs in most European countries: the project begins with the determination of the level of performance that the building must meet and, hence, the construction systems that have the potential to meet these requirements are selected.

NBR 15575 (2013) considers that the establishment of the performance level should be compatible with the protection of privacy against the intrusion of noise from activities in adjacent environments, such as speech, music, etc. The quantitative parameters of sound insulation take into account the degree of intelligibility, which is the ability to understand what is being said in an adjacent environment.

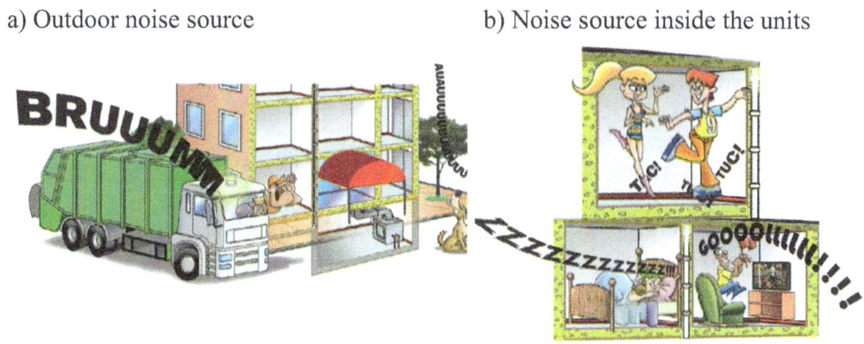

Fig. 2 Noise sources

Level of privacy and level of acoustic comfort are parameters created by Neto (2009), as mentioned by Neto et al. (2010), clarifying that these parameters may be different from the acoustic performance that a wall can present, however the authors argue that from good performance one can aim for this privacy and comfort, provided that the acoustic conditions of the receiving environment are satisfactory. Thus, it is intended that the minimum level of performance required is capable of at least blocking intelligibility in adjacent environments, ensuring such privacy.

NBR 15575 (2013) presents a table of speech intelligibility values, which were simplified from calculations found in the acoustic technical literature, due to acoustic isolation and noise level in the environment, supported by studies such as Neto's (2009). Table 4 shows the values presented by the performance standard, relating the degree of sound isolation required between environments with speech intelligibility in the adjacent environment.

Testing Methods for Acoustic Performance Evaluation and Regulatory Criteria
NBR 15575 (2013) establishes that the evaluation of the acoustic performance requirement of horizontal partitions must be carried out through specific field tests after completion of the work. For the design phase of the projects, the standard points to laboratory tests as a reference tool of the acoustic behavior of components and constructive elements individually, values that serve as the basis for the calculation of the overall acoustic behavior of the system, through a prediction method, treated in European normative ISO 12354 (2017a, b, c).

Through field trials, the verified criteria assess the actual situation of the ready-made building in terms of performance; already through laboratory tests, the results found serve as beacon values to designers, for the potential attendance to acoustic performance in the implementations of the enterprises.

NBR 15575 (2013) points out that the dormitories of the housing unit should be evaluated and establishes the following testing methods for such evaluation:

1. Engineering Method—In-ield Check—Aerial Sound and Impact Noise:

   Method performed in the field, described in NBR ISO 16283-1 (2017) and NBR ISO 16283-2 (2018), directly and rigorously characterizes the acoustic behavior of the system, both the external closure, as well as the internal walls, as well as the flooring systems, providing insulation data to the aerial sound of facades,

Table 4 Influence of acoustic insulation—DnT, w on speech intelligibility for indoor noise around 35 dB to 40 dB

Intelligibility/ability to understand what is being said aloud in the adjacent enclosure	Sound insulation $D_{nT,w}$ (dB)
Clearly audible: hears and understands	35
Audible hears, understands with difficulty	40
Audible: don't understand	45
Not audible	≥50

walls and floors between autonomous units and between a unit and common areas, in addition to the data of the standard impact sound pressure level in flooring system between autonomous units.
For the measurement of aerial sound, the methodology consists of the emission of noise in one of the enclosures through an omnidirectional sound source and measurement of sound pressure levels in frequency bands in the transmitter and receiver environments, allowing the calculation of the difference of levels with acoustic corrections of the receiving enclosure, which the method defines as standardized level (D_{nT}), which is converted into a single number through ISO 717-1 (2013), obtaining what the standard conceptualizes as a standardized weighted level difference ($D_{nT,W}$). For the measurement of impact noise, the method consists of the emission of impact noise by means of a standardized impact machine in the emitting enclosure (upper pavement) and measurement of the sound pressure level in frequency bands in the adjacent enclosure (lower pavement), allowing the identification of the standard impact sound pressure level (L'_{nT}), which is also converted into a single number, via ISO 717-2, at the weighted standard impact sound pressure level (L'_{nTW}).
The engineering method is the most rigorous and technically most complete. The results of the verification by this method are restricted to the measurements performed and are expressed in decibels (dB), and for comparison with the performance levels of NBR 15575 (2013) are adopted the values referring to D_{nTW} (standardized difference of weighted level) for the evaluation of airborne noise and $L'_{nT,W}$ (standardized impact noise) for evaluation of standard impact noise;

2. Simplified Method—In-field Check—Aerial sound and impact noise:

   Method performed in the field, described in ISO 10052 (2004), less precise than engineering, because it is done by estimating the overall sound insulation of air noise and the level of standard impact sound pressure, both in relation to the external partitions, as well as the insulation between internal environments and interfloors.
   The accuracy of the simplified method is lower, generating greater uncertainties in the results and may conflict when evaluating the attendance to NBR 15575 (ProAcústica 2017a, b). This method is indicated by the standard when there is no availability of the instruments necessary to measure reverberation time or when the background noise conditions do not allow to obtain this parameter. The procedure largely follows the engineering method with the difference that the reverberation time of the receiving space is not used (Takahashi 2016), considering this time T from tables given by the standard and the reference reverberation time T_0 as 0.5 s for the calculations. Through mathematical expression (Eqs. 2 and 3, cited by Michalski 2011), it is estimated the overall bedroom of the system from measurements 2 m from the environment, where one does not have the measurement of reverberation time of the receiving environment. First, the difference in average sound pressure level is taken from the outside of the building at 2.0 m from the façade ($L_{1.2m}$) and the average sound pressure

level in the receiving room (L_2). Taking the reference value of the reverberation time (T_0) and the calculated difference (D_{2m}), the standardized level difference is found, considering the reverberation time in the reception environment (T) estimated by the tables of the cited standard. The results of the verification by this method are also restricted to the measurements made, are expressed in decibels (dB), and the symbol $D_{2m,nTW}$ is adopted since the measurements are made 2.0 m from the analyzed element, as cited.

$$D_{2m} = L_{1(2m)} - L_2 \tag{2}$$

$$D_{2m,nT} = D_{2m} + 10Log\frac{T}{T_0} \tag{3}$$

where $L_{l(2m)}$ is the level of sound pressure in the space emitter; L_2 is the sound pressure level on the receiver espeço; T is the reverberation time of the estimated receiver space; and T_0 is the reference reverb time.

3. Precision Method—Laboratory Test—Aerial Sound:

   Precise method, carried out in the laboratory, described in ISO 10140-2 (2010), ISO 717-1 (2013) and ISO 3382-2 (2008) that provides sound insulation reference values of components and building elements, such as wall, windows, doors, etc., calculated to serve as a basis for project development. First, each component is tested and then the overall bedroom of the construction system is calculated. It is the most accurate method to achieve the sound reduction index, *R*.
   The measurements performed in the laboratory are performed in two reverberant chambers, one of emission and the other receiving, constructed and qualified for laboratory tests, as characterized by ISO 140-1 (Takahashi 2016). The results obtained in this method are expressed in decibels (dB), adopting for their results the symbol R_w.

Air sound insulation values measured in the field mean a standardized difference in weighted level of the checked system, whether between environments, external partition or coverage. They are expressed by the D_{nTW} or $D_{2m,nTW}$ convention. The indexes measured in laboratory tests indicate the weighted sound reduction of the constructive component used, providing calculation reference values for the projects, thus estimating that there will be potential to meet these values when tested in the field. These are expressed by the R_w. Convention For impact noises, the variable $L'_{nT,W}$, indicates the weighted standardized impact sound pressure level.

With regard to the criteria for evaluating the acoustic behavior of a building, i.e., the expected quantitative indices (values), the NBR 15575 (2013) standard determines the following factors:

1. Sound insulation of walls between environments—criteria for in-field tests;
2. Sound insulation of walls between environments—criteria for laboratory tests;
3. Insulation of facades and roofing—criteria for in-field tests;
4. Insulation of facades—criteria for laboratory tests;

5. Air noise insulation between floors and accessible covers;
6. Insulation of impact noisefrom accessible floors and covers;
7. Noise insulation caused by hydrosanitary equipment.

NBR 15575 (2013) translates these criteria as required minimum values, that is, it requires the guarantee of the minimum, although it brings a minimum to the higher level classification, passing through the intermediate performance condition, where for each type of environment are determined ranges of values within this classification. The tables presented and detailed below (see Tables 6, 7, 8, 9, 10, 11, 12, 13 and 14, except 8) bring these ranges of values for each situation evaluated.

For the acoustic attenuation of noise between airborne floors, the standard presents limit values of the difference between the standard noise generated in the autonomous unit and the sound intensity recorded in the evaluated unit. Table 11 indicates the limits, the values of which are the result of field trials.

According to the CBIC (2013) guide, the level of performance in relation to air noise increases as the mass of the element is higher, in compliance with the Law of the Masses. An example is given for concrete slabs of 10 cm thickness, whose acoustic insulation corresponds on average to 45 dB, taking into account the minimum level of performance, reaching the top level of performance when the thickness increases to 15 cm. On the other hand, the noise produced from the impact of solid body, the standard determines that the sound pressure level of the impact standardized by the ISO 140-7 method in the analyzed environment, from the field test, should not exceed the limits presented in Table 5.

As opposed to the relationship of air noise with the Law of The Masses, the bedroom to impact noise is impaired the denser the material, as there is greater acoustic transmission directly by the dense element. In the example given by the CBIC (2013), the same concrete slab of thickness from 10 cm to 12 cm is located on the margins of the minimum level of performance. As a suggestion to increase this performance, the CBIC indicates the use of light concrete, acoustic or suspended liners, when it is not possible to perform floating floors, a solution of better acoustic absorption that will act as a mass/spring/mass system, enabling compliance with better performance levels.

Table 5 Weighted standardized impact sound pressure level, $L'_{nT,w}$

Element	$L'_{nT,w}$ (dB)	Performance Level*
Floor system by separateing autonomous housing units positioned on distinct floors	66 a 80	M
	56 a 65	I
	≤55	S
Accessible coverage or floor system of areas for collective use (leisure and sports activities such as home theater, gyms, ballroom, games room, collective bathrooms and changing rooms, kitchens and collective laundries) on autonomous housing units	51 a 55	M
	46 a 50	I
	≤45	S

(*) M: minimum level/I: intermediate level/S: Superiortop level

For noise bedroom caused by hydrosanitary equipment, although not mandatory, the standard indicates maximum values of the perceived sound pressure level during an operation cycle of the equipment that emits noise, $L_{Aeq,nT}$, and maximum values during operation, $L_{ASmax.,nT}$. These are noises produced in collective water and sewage pumps, discharge valves and other equipment fired in neighboring units and are indicated in Tables 6 and 7 (CBIC 2013). Such analysis is not part of the scope of this work.

For the in-field tests, NBR 15575 (2013) determines not only the evaluation methodology, but also the types of equipment and recommendations for the use, calibration and handling of equipment. Almeida et al. (2006), presented in their study the description of the main equipment for measuring field acoustics:

- Sound monometers: sound pressure measuring equipment, which uses various forms of units, through the use of a measuring chain consisting of a microphone (which converts sound pressure into an electrical signal), a signal amplifier, a frequency weighting circuit, an effective values circuit and a dial;
- Frequency analyzers: equipment that allows you to analyze signals in the field of frequency a;
- Dosimeters: apparatus that allows a worker's exposure to noise to be assessed, providing the analysis of average noise values as well as short-term peaks;
- Sound intensity probes: devices that allow the sound intensity to be evaluated in a given direction, caused by a source.

In addition to these mentioned equipment, the standard also provides for a sound pressure level meter calibrator, which next to the meter (sonometer) must be certified every 2 years by the Brazilian Calibration Network or the National Institute of Metrology, Standardization and Industrial Quality (INMETRO). Another instrument used is the noise-emitting source, which is an equipment used to reproduce noise with sufficient intensity to measure the transmissions from one environment to another, thus allowing the acoustic bedroom of the paraments. Figure 3 exemplify some of these measurement equipments used.

Table 6 Maximum values of the equivalent continuous sound pressure level, $L_{Aeq,nT}$, measured in bedrooms

$L_{Aeq,nT}$ dB (A)	Performance level
≤30	S
≤4	I
≤37	M

Table 7 Maximum values of the maximum sound pressure level, $L_{ASmax,nT}$, measured in bedrooms

$L_{ASmax,nT}$ dB (A)	Performance level
≤36	S
≤39	I
≤42	M

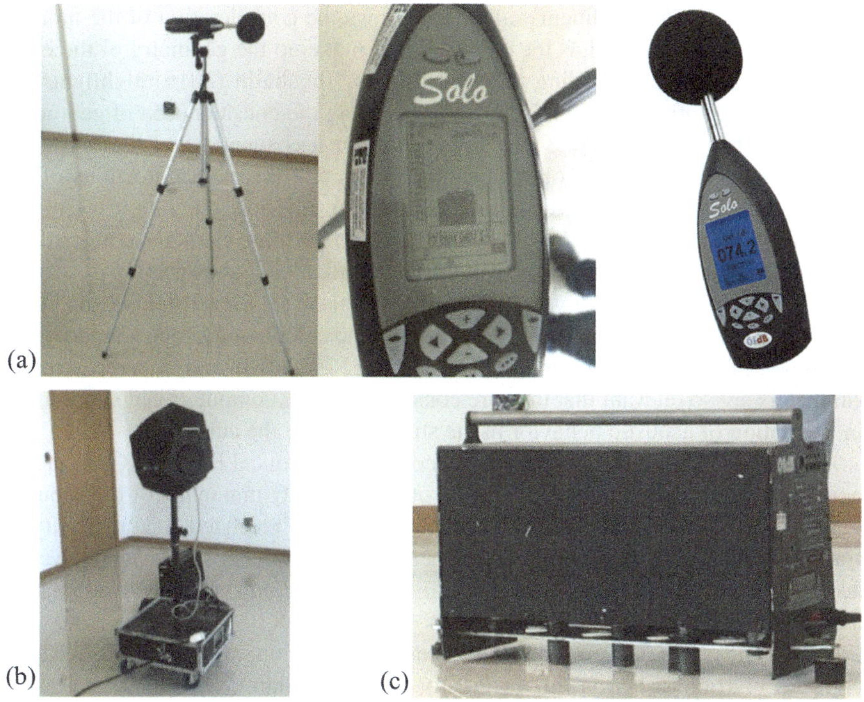

Fig. 3 **a** Sleepmeters; **b** Noise source; and **c** Impact noise production equipment (Tapping Machine)

Acoustic Performance Prediction Method—Numerical Simulation The numerical simulation, although not foreseen in NBR 15575 (2013), since this standard is not prescriptive and presents only the evaluation methodology by tests, is the method of calculation of acoustic performance prediction based on European standards EN 12354-1,2,3 (2002) (air sound insulation between environments, sound insulation of impact between environments and air sound insulation against external noise, respectively) that was internationally standardized by ISO 12354-1,2,3 (2017), and which allows, therefore, the estimated calculation of acoustic insulation between environments, in order to simulate the results of field measurements predicted at ISO 16283-1 (2017), ISO 16283-2 (2018) and ISO 717-1 (2013). These numerical simulations allow the evaluation of such acoustic criteria in accordance with NBR 15575 (2013), even before the works are executed.

Computational acoustic simulation programs are important tools for calculating performance prediction models (Takahashi 2016). In his work, Takahashi (2016) points out that the European standard ISO 12354 (2017), in parts 1, 2 and 3, specifies the calculation models for calculating the sound insulation of facades and internal partitions, through the use of the equations foreseen and detailed in the standards that guide the field tests (ISO 140-4, ISO 140-5, ISO 717-1), and these consider, as parameters in the calculations, in addition to the direct transmission of noise, the flanking (slabs and/or confluent walls) that allows the consideration of the indirect sound contribution, as well as the elements that make up the geometry of the environments or facades. Regarding these conditions, Takahashi (2016) highlights the possibility of positive and negative effects regarding the consideration of geometry, that is, the shape of the façade may imply more or less sound transmission due to the total or partial shielding of the façade casings and the extra reflections of the ceilings of the balconies. As for indirect sound transmission, the author highlights the transmission via flanking very common in rigid building elements connected to other rigid elements, a common condition of Brazilian civil construction due to the use of concrete and brick in walls and slabs. In this case, the author considers the need to consider the contribution by flanking in the calculation of the sound contribution, even though it is not discussed by NBR15575 (2013), and in this case these parameters are so relevant that they are considered in the computational simulations for prediction of acoustic behavior in the study made by the authors.

As the European standard takes into account the Technical Building Code of that region, the CTE, Takahashi (2016) points out in his study that many of the acoustic performance simulation programs developed were not idealized for the reality of Brazilian performance regulations, but because they establish similar objectives to meet the performance criteria of the buildings, are used and studied with adaptation to the Brazilian market.

For the most part, the programs were developed in Europe and the USA, whose reality of civil construction differs greatly from Brazilian civil construction, and, although they have a vast database of types of construction elements, due to specificities and constructive differences, require greater work by researchers, who need to seek information from laboratory tests to obtain data and technical data from manufacturers and construction elements in order to be able to insert such data with quality and reliability in the programs and, hence, perform their simulations, thus adapting the software to the reality of Brazilian civil construction. According to Carvalho (2015), the reason for the scarcity of prediction tools developed by Brazil is the little interest of research and development in this subarea, which is why the vast majority of the tools available in the market are imported.

Some programs have been more cited and used in research such as SONarchitect ISO, INSUL and Bastian, with recognized efficiency in studies on the acoustics of buildings, but other programs such as Projetus of the company Multinova, CadnaR and Cadna-A of Datakustik GMBH are also acoustic forecasting programs that provide specific planning, noise mapping or modeling according to calculation procedures based on European standards ISO 12354-1 (2017a) and ISO 12354-3 (2017a), or in the environmental noise recommendations of Directive 2002/49/EC (European Environmental Noise Directive, standardised by ISO 17534). However, each program has its limitations and particularities.

Due to the low variety of such programs, it is observed that the researchers end up adopting the available ones and adapt them according to the desired objectives in their studies, as was the case of the Giunta (2013) research that used the Cadna-A program for the modeling of acoustic map of part of the city of São Carlos-SP, Carvalho (2015) that used the Programs SONarchitect ISO and Insul to analyze the levels of acoustic performance to impact noise in works of São Paulo Paulo-SP, Ferreira and Zannin (2007) who used the Bastian program to evaluate acoustic transmission in residences in Curitiba and Santos Filho et al. (2017) who used the Insul program to simulate the acoustic performance of ventilated façade systems in Brasilia.

In his study, Giunta (2013) used the Cadna-A program, which among the programs developed for the generation of noise maps, this was, at the time, one of those found on the market that used the French model NMPB 2008, a model of noise calculation produced by road traffic recommended by the European Directive 2002/49/EC (OJEC 2002). The author verified the applicability of this same model to the Brazilian case in her study, comparing with the mapping performed by the Geographic Information System—GIS.

Considering that the sound mapping of an urban area is a tool of great importance in urban and territorial planning, and that among the purposes are the identification of points or critical areas regarding environmental noise, Giunta (2013) points out that performing sound mappings through on-site measurements or even through manual mathematical calculations, are time-consuming and very expensive processes and therefore, it studies how the models created by computer programs can simulate the urban acoustic environment from these same calculations.

After the simulations, Giunta (2013) was able to conclude that the simulation allows identifying critical points of noise flow, future problems and thus enables prevention actions, whether external protection to buildings, or adjustment in arrangement of environments and openings for less noisy directions. It also concluded that the preliminary simulation with gIS proved to be positive as a simplified mapping for evaluation of large areas, however this does not reproduce local specificities, presenting values above real noise levels.

With Cadna-A, although the use of the program is more laborious and time-consuming, due to the possibility of including richness of details, it was possible to analyze the three-dimensionality of the studied area, enabling the analysis of sound propagation in the horizontal plane, cuts and perspectives (Giunta 2013). Even so, like the other programs developed outside Brazil, the author highlighted that some fields of the program would need to be improved for better approximation with real data and also application to Brazilian civil construction, such as the inclusion of technical data on the types of Brazilian pavements, in addition to local weather conditions. Generally speaking, Giunta (2013) highlighted that although the simulation with Cadna-A indicated some low correlation values, the deviations presented were considered within acceptable limits by the research.

Studying the impact noise in slabs, Carvalho (2015) used the Insul program, together with the SONarchitect ISO program, in a comparison of acoustic performance between simulated slabs based on the result of the acoustic performance field measurements of these slabs, as the initial reference. The use of these two programs is justified by the possibility of creating computational models analogous to those of field environments and adopting the same performance classification according to the standardized methodology for field tests, which are ISO 12354 standards. To create computational models, the author used Insul to model the construction elements and extracts the acoustic performance input data from the slabs and walls, and modeled in the SONarchitect ISO program the same ambient volumes of the samples measured in the field, also inserting the performance classification indexes of the Brazilian standard NBR 15575 (2013). In his study, Carvalho (2015) was able to perceive a linearity in the prediction results, even when the difference between the results reached 6 dB in relation to the one measured in the field in his samples, and highlighted that this difference is due to the fact that the simulation takes into account a perfect model, where there are no flaws in structural junctions of the building and also where the masses and densities of the building elements are always controlled, a situation that cannot be guaranteed in a real work.

Another study, conducted by Santos Filho et al. (2017), recorded that the Insul program, although it was used to predict the bedroom of ventilated facades, had as a limitation the impossibility of considering all the elements of the façade assembly simultaneously in the calculation of the prediction of sound insulation. Thus, the systems were simulated separated from the frames, and later they used an equation weighting these values calculated separately. Even so, the authors concluded as positive the simulation, in the sense that the simulation presented values close to those found in the field measurement tests, thus allowing designers to provide better insulation indexes to the buildings, as I can test the types of elements and compositions during the design phase.

Another widely used program (Ferreira and Zannin 2007; Ferreira 2004), Bastian, from DataKustik, has a database of construction elements measured in a laboratory in Germany, and as well as other programs developed outside Brazil, the indexes are quite different from those used by Brazilian standards.

Ferreira and Zannin (2007) used Bastian software to determine the coefficients of sound insulation in facades and between residential rooms, comparing the results measured in field tests and simulated through computer modeling programs, and had to look for information from laboratory tests of the building elements of Brazilian civil construction, specifically the systems adopted in the 11 residences studied. They pointed out that, in addition to brazil's climate being very different from Europe, the constructive pattern and purchasing power of that region, suggests greater demands for acoustic comfort, which is why the specification of systems and elements are very divergent. However, even adapting the Bastian program to serve them, the authors cited that the Bastian has the limitation of modeling the geometry of the environments, that is, the software only allows the inclusion of rectangular geometries, forcing the researchers to adjust the geometry of the studied environment to this condition, while maintaining the same volume and the same area of the wall that would be exposed to noise. Thus, the use of Bastian implies simplification of the architectural form of the environment for the use of the tool, and with this simplification, the reverberation conditions differ from the actual field situation. In this sense, there is a risk of simulating environments with architectural characteristics distinct from the real situation, which can significantly influence the acoustic result of the environment, because there is a great relationship between architectural characteristics and the sound quality of the environments, as mentioned in a study by Takahashi and Bertoli (2012). The work of the authors Ferreira and Zannin (2007) concluded that measurements are extremely important, and that simulations require expensive equipment and well-trained staff so that the results express a greater approximation to the real performance of the environments, in addition to adapting to the reality of the typology of materials and building elements of Brazilian civil construction, because often some of the input data cannot be adjusted and simulates with European or U.S. standards, which the authors pointed out as extremely worrying.

Some studies, however, have been developed for new software according to the criteria and parameters required by NBR 15575 (2013), such as the example of the thesis presented by Takahashi (2016) where the Acoustic Hipnos program is developed and presented, which makes it possible to simulate with local systematized data to specifically meet NBR 15575 (2013). Although it is a creation of the Brazilian research market, the program has some limitations, due to very little information of acoustic performance data of the construction materials, in addition to the limitation of these data in the form of unique value for the system studied and not as a function of frequency, causing the system not to reproduce the situation of the walls composed of more than one material. And, being the result of a relatively recent thesis, it is still intended to evaluate the difficulties presented in the use of the program, to later think about spreading for commercial use.

In this work, the Programs SONarchitect ISO, sound of numbers and INSUL of Marshall Day were used. The choice was due to the use of programs based on international regulations ISO 12354, because they were used by other researchers in similar studies, because they are easy to use, because they allow the modeling of the environments according to the faithful geometry of the projects, because they already have more than 15 years of development and research, and also by the availability with the company specialized in acoustic performance projects of the local market, which supported the realization of the present study. However, in any of the programs used, the quality of the input data of technical and physical characterization of the elements and environments, such as the surface densities of the materials and elements, the thicknesses of the layers, the volume and shape of the environments and others, must be judicious and correctly informed, because the incorrect specification of any of these data may compromise the results calculated in these tools and, therefore, estimate performance values far from the actual performance indices measured in the field. Moreover, as Takahashi (2016) quoted, a building built according to the materials indicated after the computer simulations, but with poor execution, does not guarantee the achievement of the projected and desired acoustic performance, which therefore requires a supervision of the quality of the execution of the work to achieve the performance provided for in the prediction tool.

Difficulties using the Standard NBR 15575 (2013) More comprehensively in relation to meeting the regulatory requirements of NBR 15575 (2013), Cotta and Andery (2018) analyzed how project companies, specifically studying 4 companies operating in the region of Minas Gerais, Espirito Santo, Rio de Janeiro and Brasília) have been appropriated and using innovative tools to raise awareness of the importance of meeting such requirements, showing, however, that the difficulties associated with technical solutions and the lack of integration between disciplines remain the main obstacles to full compliance with the performance standard.

Specifically in the acoustic performance requirement, some studies have been done to demonstrate that, even with explanatory publications to help comply with the regulations, many obstacles still corroborate the low compliance with the minimum criteria of the standard, making the acoustic adequacy of the buildings neglected since the project stage.

It is noteworthy that these obstacles are related to the time when the norm is in force, relatively recent, which makes considering such arguments justifying that those involved, whether builders, designers or manufacturers of materials, are still in the phase of adaptation and reorganization of their processes. The speed of the process of design and execution of the projects, combined with cost reduction, caused acoustic performance to be put in the background (Resende et al. 2014). On the other hand, this is the most "felt" requirement by the user.

Building construction in Brazil, for years, has underestimated the issue of sound insulation in the project phase of the projects and, therefore, the frequent problems of execution impair the sound bedroom of the construction elements, potentiating the low bedroom between the housing units. According to a study by Catai et al. (2006), the main partition elements used in Brazilian civil construction, such as ceramic blocks, concrete blocks and reinforced concrete itself have some sound insulation, however it does not always happen satisfactorily, due to the execution failures in the construction process. [...] Only with the obligation to comply with the normative criteria, especially due to the legal consequences of non-attendance, is that civil construction returned its gaze to acoustics. Still, many barriers have been faced in the implementation of NBR 15575 (2013).

Morais et al. (2018) studied the main obstacles faced by the designers of Pernambuco and Alagoas in the adequacy of their projects to the acoustic criteria of the performance standard and demonstrated, through the research, that the lack of consideration of the surrounding conditions, the lack of knowledge in relation to the insulation capacity of the materials, and the high cost linked to the project with the acquisition of specific materials collaborate with the low compliance with the acoustic requirements of the said standard.

The research by Silva et al. (2014) in offices in Porto Alegre, demonstrates that, to adapt the projects to the acoustic requirements of NBR 15575 (2013), in addition to better technical knowledge of the materials and innovative acoustic solutions in the market, designers need longer deadlines, exchange of experiences between the category and training of professionals.

Moreira et al. (2018) studied how the construction companies in Fortaleza have been adequate to the acoustic performance requirements, identifying the difficulties faced, the changes and strategies used in terms of design and specification of systems to meet the minimum normative criteria. They concluded that there is a divergence of solutions due to the lack of knowledge of the new acoustic insulation materials available in the market, choosing the high cost as a limiting factor; and the choice of minimum performance as an alternative to better balance the cost/benefit of your enterprises.

This limiting cost issue related to compliance with minimum levels of acoustic performance coincides with what Boufleur (2013) says that the lack of sound insulation "is caused by neglect with the importance of acoustic performance, when choosing the noise attenuation system, being often analyzed only by economic aspects and practicality of execution", and the consequence of the economy is the failure to comply with the necessary bedroom after execution, especially if the constructive and executive details or those caused by execution failures compromise the minimum solutions designed.

For the researcher Paixão (1997), what existed in general was a mistaken idea that providing comfort would imply a significant increase in cost to the enterprise, exactly because of ignorance of the technologies offered, however what came to understand is that the costs and impacts to correct the problems arising in the post occupation, especially with regard to noise control are so much more significant that they end up making many of the possible solutions unfeasible.

Hence the importance of adopting an ideal path to comply with the acoustic performance parameters in the project phase, where the theme is prioritized, seeking noise control solutions before performing the work, in order to minimize post-work adaptation costs and ensure the minimum compliance with the necessary acoustic insulation.

In resume, this work has as general objective to analyse the efficiency of computational simulations of acoustic performance of the building and its systems, currently used in the design phase of housing developments, through the use of the Programs SONarchitect ISO and INSUL.

The specific objectives of this work are:

- Collect data from field tests carried out in 14 residential developments, objects of the case study, for horizontal partitions;
- Perform computational simulations of the acoustic performance expected in the in-field of these enterprises raised in horizontal partition systems (SVH);
- Compare the values of the in-field tests, with the values resulting from the computational simulations, whose input data of the components are based on the project specification sheets and compatible with data results of previous laboratory tests. Of these 228 comparisons, 116 results for air noise in horizontal partition systems (SVH-RA) and 102 results regarding impact noise between floors of residential buildings (SVH-RI);
- Analyse the comparative data found, i.e., measured values versus simulated values;
- Evaluate the influence of modelling procedures on the assertiveness of the values found in the simulations;
- Evaluate whether computational simulation represents reliability for the acoustic performance criteria required normatively, based on the results of the case studies used.

2 Comparative Evaluation—Methodology

2.1 General Aspects

In this work of comparative analysis of computational simulations and in-field measurements was performed, i.e., the acoustic behaviors of projects whose field measurements had already been performed in the delivery phase of these works were simulated. With the simulation in the post-work phase, it was intended to verify whether the simulated performance values were compatible with the performance results measured in the field.

Thus, the compatibility between the measured and simulated acoustic performance values was evaluated, based on the data contained in the in-field test reports and the data calculated after the result of the computational modeling of the selected sample properties.

As the scope of this work is restricted to the analysis of sound insulation of horizontal partition systems (SVH), only the data of:

a. air noise insulation through horizontal partitions (SVH-RA);
b. sound pressure transmission from floor impact noise (SVH-RI);

not including in the present study the analysis of sound insulation data of external vertical partitions (SVVE-facades), although in the field reports of the properties contained such results.

The methodology applied for the comparative evaluation of field data with the data from the computational simulations was developed following 8 steps listed below:

1. selection of residential properties with measurements made, called the selection phase of the case-studies;
2. tabulation of the values obtained in the field tests;
3. analysis of the results of field trials;
4. preparation of architectural design for simulations, through AutoCAD software;
5. composition of materials and building elements in specific soundproofing software—Insul 9.0;
6. modeling the building and calculating the overall sound reduction through the computational simulation of the building as a whole through soundproof calculation software—Sonarchitect ISO Professional;
7. tabulation and analysis of the results obtained in the computer simulation;
8. comparative evaluation of field values versus simulation values.

2.2 Experimental Campaign and Numerical Simulation

2.2.1 Selection of the Case Studies

It was selected 14 residential buildings to analyse in detail. In 5 cases the building floor accompanied the type of construction system, using solid concrete slabs with a thickness of 10 cm; the other buildings used ribbed slab, also called alveolar slab, varying the thickness of solid concrete covers between 5 cm and 7 cm.

Acronyms were adopted to identify the type of construction system, for example, for the ribbed slabs the acronym adopted was LN and for the solid slabs, LM. Table 8 shows the quantity of each type of wall masonry and type of slab in relation to the total sample analyzed.

The buildins analysed, all from the Northeast region, are located in the cities of Maceió-AL, Camaçari-BA, Fortaleza-CE, João Pessoa-PB, Cabo de Santo Agostinho-PE, Paulista-PE, Recife-PE, São Lourenco da Mata-PE and Aracaju-SE.

The field tests were carried out between the years 2017 and 2018, in the delivery phase of these works, and the company that executed them used equipment standardized and calibrated by laboratories accredited by Inmetro, adopting the recommendations of the standard regarding the procedures of the tests. The specifications of said equipment used in field tests are set out in Table 9, including calibration certificates, were within the period of validity at the time of the measurements (two years from calibration) and in each test report the copy of the respective certificates was attached.

From the in-field test reports, data from the following variables were evaluated:

Table 8 Total samples of the case study

Floor	Nr.
LN- Ribbed slab	9
LM—concrete solid slab	5

Table 9 Equipment used in-field tests

Equipament	Manufacturer	Serial No./version	Calibration certificate
Sound Pressure Level Meter (Sleep gauge)—Class 1	01 dB	65279	RBC2-9646-555
Acoustic Calibrator—Class 1	01 dB	34113649	RBC3-9649-666
Dodecahedron (Omni-12)	01 dB	03/09-12/B206-012	–
Amplifier	01 dB	03/09-12/B207-A12	–
Tapping Machine	01 dB	CALP04/08-11/193	–
Software dBBati (compilation of the results)	01 dB	–	–

Table 10 Sample Summary, type of construction system and acoustic tests

Building	Location	Code	Floor	In-field	
				SVH-RA	SVH-RI
I-1	Recife/PE	LN-01	Ribbed slab	8	4
I-2	Recife/PE	LN-02	Ribbed slab	10	5
I-3	Aracaju	LN-03	Ribbed slab	7	7
I-4	João Pessoa	LN-04	Ribbed slab	4	4
I-5	Recife/PE	LN-05	Ribbed slab	3	3
I-6	Recife/PE	LN-06	Ribbed slab	20	20
I-7	João Pessoa/PB	LN-07	Ribbed slab	10	10
I-8	Fortaleza	LN-08	Ribbed slab	8	8
I-9	Cabo Sant Agostinho	LN-09	Ribbed slab	8	8
I-10	Maceió	LM-01	Concrete solid slab	8	8
I-11	Camaçari	LM-02	Concrete solid slab	8	8
I-12	Camaçari	LM-03	Concrete solid slab	8	8
I-13	S.Lourenço Mata	LM-04	Concrete solid slab	6	6
I-14	Paulista	LM-05	Concrete solid slab	8	3
Total				**116**	**102**

- SVH-RA: Sound insulation to overhead noise promoted by horizontal partitions (floor system) = 116 results;
- SVH-RI: Impact noise in floor systems between units (horizontal partitions) = 102 results.

Table 10 presents the summarized description of the sample characterization, with data from the city of location of the enterprise, type of construction system of the walls and slabs, and quantity of field tests performed for each type of variable for each study case. The measurement procedures of the field tests followed the methodologies standardized by ISO 10140-2 (2010) and 16283-1 (2017), with a minimum of five measurements in each environment, with the doors and windows well partitioned and the environment empty, except in the I-5 building where there was already some furniture installed by the owner at the time of the test. Figure 4 illustrates the measurement scans of the field tests for the three variables of the evaluated partition systems (SVH—air borne noise and SVH—impact noise).

2.2.2 Numerical Simulation

The Marshall Day Acoustics' Insul 9.0 software was used as a soundproofing device for the elements of walls, floors and ceilings, in bedroom. This initial modeling of the components, provides the predicted values of sound reduction of the elements, which will be reported as input data then in the overall simulation of the building.

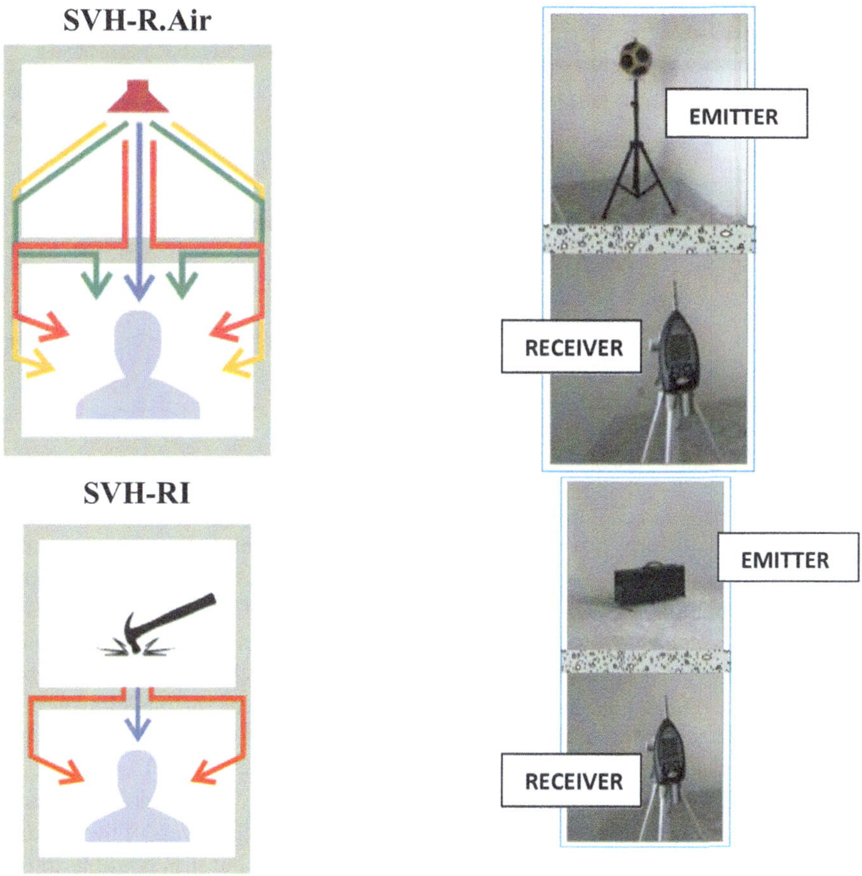

Fig. 4 Schemes of the in-field test measurements

According to the manufacturer, Insul does not replace the measurement, however, as a prediction tool, it makes good estimates of sound transmission loss and impact sounds in bands of one-third octave and weighted sound reduction index, with accuracy and reliability within 3 dB for most types of constructions, with respect to the values measured in the field. With the program, it was also possible to evaluate the effect of material change in the projects, through adjustments in the modeling, considering, for each material, the different densities, modulus of elasticity and specific damping factor. The manufacturer further clarifies that the program is multilingual and has a database of materials from various countries of the world, such as UK, USA, Australia, Netherlands, Spain, France, Germany, Italy, Sweden, Korea and others, which can be edited if necessary. Figure 5 shows the Insul screen, which contains part of the settings and frequencies of the acoustic calculation considered by the program.

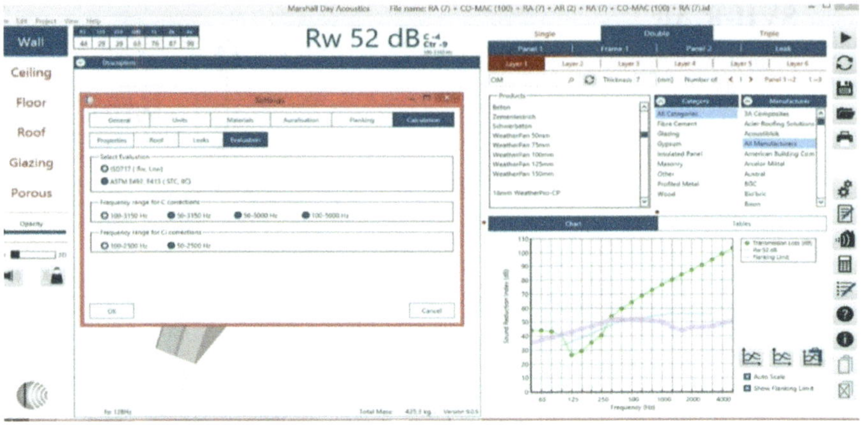

Fig. 5 Configuration screen of standardized calculations in Insul Software

With Insul software, it was possible to simulate the sound attenuation possible by the components and building elements (walls and floors), to reach the reference values of insulation R_W and $L_{n'W}$ for each composition. The insulation values of the materials calculated by Insul, R_W and $L_{n'W}$ were taken to the other computer simulation program used in this work, the Sonarchitect, which simulates the acoustic behavior of the building as a whole, evaluating its performance. The modeling in Insul was structured according to the densities and thicknesses of the different layers that make up the wall systems and floor systems.

As previously mentioned, the program has a database with characteristics of various types of materials to be used in the compositions of the layers of the building elements to be modeled, in this case the walls and floors, but it is possible to edit or create custom compositions. In this screen of editing and creation of new material, the information of layer thickness, density, young modulus and damping were inserted for each of the layers that make up the building elements of each project to simulate.

In this work, all compositions were created in a personalized way, according to information from the project specification sheets of each project, which contain the characterization of the flooring systems used.

As there are no technical publications detailing the criteria for the use of the acoustic damping factor of the materials and, as the densities of most materials are known, in addition to reference values of sound reduction of some elements and components, which were the results of laboratory tests provided by the tables of the CBIC (2013) guide, data of other researches such as Neto (2009), Neto et al. (2010) and Souza et al. (2013) and the FAD Performance Evaluation Forms of the Brazilian Habitat Quality and Productivity Program—PBQP-H, we chose to adjust the damping values used in the simulations of this study, in order to reach these acoustic performance reference values of other studies.

The information related to the properties of density, modulus of elasticity and damping factor of the materials used in the simulations of the case studies were provided by Tecomat, based on the field tests and the data of the manufacturers of the materials used in the works, compatible with the data adjusted above and are listed in Table 11.

The models of the walls in the Insul program were performed as the masonry with its coating are executed in the field, that is, as being a juxtaposition of several overlapping layers as a monolithic element, which are coating and substrate layers, which in the present work were in masonry or solid concrete, with their respective thicknesses and properties of density, modulus of elasticity and declared damping factor.

For each wall composition created, the software performed the calculations of sound attenuation showed an image of the element as constructed and indicated the sound reduction index in the frequency ranges requested, plotting graphs and tables, presenting the critical frequency and total density, in addition to the mean Rw value for the 100–3500 Hz band, which was used as a reference index of the acoustic behavior of the respective element.

For the slabs, although the layercomposition methodology is similar to those used in the modeling of the walls, the modeling was a little more complex, because

Table 11 Characterization of simulated materials

Acronym	Material	Density (Kg/m^3)	Young's modulus (GPa)	Damping
BCE VED 90	Ceramic partition block 90 mm	616	10	0,011
BCE VED 140	Ceramic partition block 140 mm	577	10	0,011
BCO VED 140	Concrete partition block 140 mm	896	40	0,001
BGV	Cast plaster block 5 mm	800	8	0,006
BGM/FG	Solid plaster block/Plaster lining	900	8	0,006
CON	Solid concrete wall	1900*	15*	0,003*
RA	Cementmortar coating	1600	30	0,003
RG	Plasterpaste coating	1100	30	0,003
RC	Ceramic coating	1800	4,68	0,001
CP	Counterfloor	1600	30	0,003
LJ	Concrete slab	2400	40	0,009**
CPE	Special floor—trace isofloc	1100	10	0,004*
MAF	Asphalt blanket with acrylic resin			

a weighting of the different materials of the floor layers was made for the purpose of commaking the slab as a mass, that is, a single equivalent component, calculating a weighted density of the set, according to the specific thicknesses and densities of each material per layer, from the following expression:

$$\rho' = \left(e^1.\rho^1 + e^2.\rho^2 + \cdots + e^n \cdot \rho^n\right) / e^T \tag{4}$$

where ρ' is the total density expressed in kg/m^3; e^n is the thickness of each layer that makes up the slab, expressed in m; ρ^n is the density of the material of each layer, expressed in kg/m^3 and e^T the total thickness of the massif, expressed in m.

This weighting that transformed the various layers of the slab into an equivalent layer was made because, otherwise, the Insul program would consider that between the layers of the floor there would be an elastic joint, and thus the simulation would present much better results than those that occur in the real situation, that is, in the field trials.

In the case of ribbed slabs, the simplification was even greater, because the height of the ribs was disregarded and, for the calculation of the equivalent massif, only the thickness of the slab capping was considered. This greater simplification can be justified by the fact that, ignoring the air space of the ribs, the process leads to lower levels of sound reduction, a solution adopted that "favors" safety.

After modeling the slab (floor/ceiling), the Insul program performed the acoustic performance calculations and indicated the Rw sound reduction index of the ceiling element and the standard impact sound pressure level of the Ln,w floor in the Ceiling and Floor tabs of the program, respectively. For each of the Ceiling and Floor tabs, the program also presents the values of air driving noises R(dB) and percussion noises Lw(dB), at frequencies from 50 to 5000 Hz, in the Tables tab, but for the modeling we use only the values included in the standard range band (band of one third octave) whose frequencies range from 100 to 3500 Hz (values that will be exported to the other modeling program—the Sonarchitect).

2.2.3 Acoustic Performance Calculation

For the acoustic performance calculation the software Sonarchitect ISO professional was used. The program was used to simulate the acoustic behavior of the building as a whole and evaluate its performance from the conformity of the simulated results with the defined sets, according to the constants in NBR 15575 (2013), which are initially loaded into the program. The results presented by the program are detailed for each environment, partition, flank and sound transmission path and is presented the conformity or not of each environment/system.

The values resulting from the simulations performed in the modeling of composition of materials and elements with the Insul software were taken to the Sonarchitect software, as reference values of the custom compositions of each of the projects, with a view to computational modeling of the acoustic performance of the building.

In addition to the composition data calculated in the Insul software, we used the floor plans exported in DXF, the information of the architectural project (between-floor height, window indication, specifications of internal, external, floor and ceiling partition) and the values of the minimum limits of the acoustic performance criteria for in-field tests, according NBR 15575-4 and NBR 15575-3 (2013), used as input data for computer simulation in Sonarchitect.

The modeling process in Sonarchitect started with the insertion of the minimum limit settings, i.e. the minimum performance levels of NBR 15575 for the field tests, since the simulation intends to predict the acoustic behavior in relation to what will be measured in real in-field situation.

As this work did not evaluate the criteria of external partition systems (SVVE), information on façade frames was not included in the program, since this information would not significantly influence the calculation of the internal transmission of noise between internal and between-floor partitions, which were the restricted criteria evaluated.

After the feeding of the building characterization data, the program calculated the acoustic insulation of the entire building, indicating, for all environments, the values of aerial acoustic transmission and impact of sound pressure levels and the components and paths of noise transmission. The program also compared the result of the acoustic simulation calculated with the values of limits reported in the initial configuration, demonstrating the environments inside or outside these minimum level limits required by NBR 15575 (2013), through green and red colors that mean, respectively whether or not they met the required by the aforementioned standard.

Figure 6 illustrates how acoustic performance behavior plots are at the end of this computational simulation phase using Sonarchitect. We are exemplified how the projects modeled in 3 dimensions are presented; how the floors clearly present, through the red color, the environments in which values are identified in conformity in relation to the minimum levels of the standard; as the program presents the acoustic analysis for each environment, with its relationship with the neighboring enclosures on the same floor and between floors; the values of simulated and standardized sound insulation; and the reports and graphs presented for each type of noise analyzed.

With the modeling of the elements, design and characterization of the project, the computational simulation of the acoustic performance of the building was made, demonstrating the acoustic behavior of the ready building, succeeding in acoustic performance results by environment, allowing the confrontation of measured values and simulated values.

From the field, sound reduction values were obtained between environments, in this case the standardized difference of weighted level of air noise of horizontal partitions (D_{nTW}) and the level of standard impact sound pressure ($L'_{nT,W}$) between floors.

From the computational simulations, first modeled in Insul, the acoustic behavior of reference of the construction elements was predicted individually, based on the projects and specifications of the construction systems, and the sound attenuation indexes in relation to air noise (Rw) and the standard impact sound pressure level for impact noises ($L_{n'W}$) were calculated.

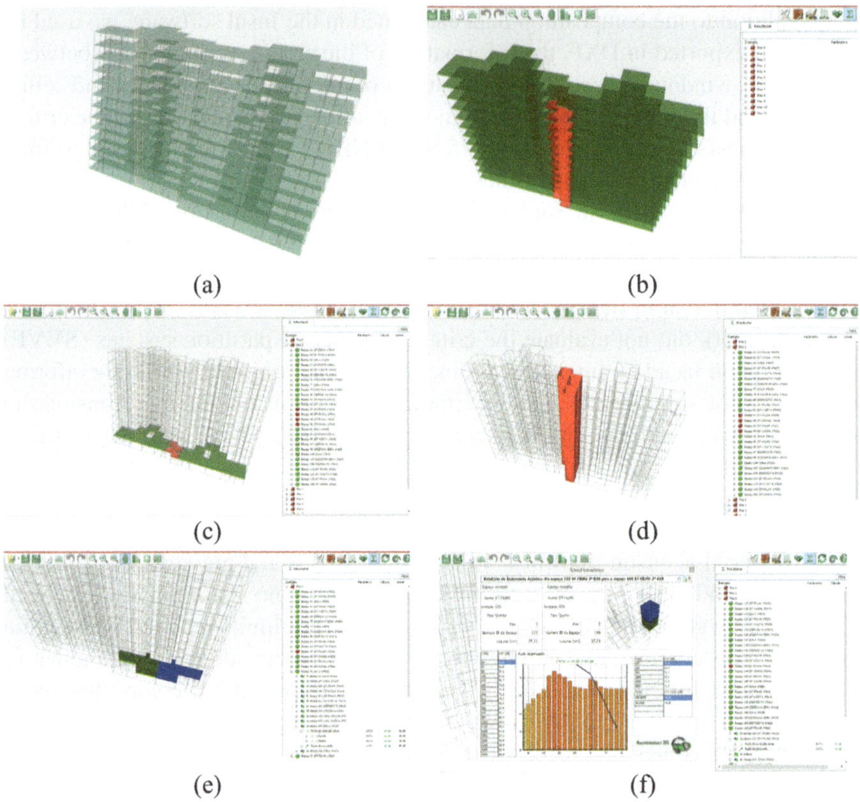

Fig. 6 **a** Example of the 3D modeled project (Building I-9); **b** Plotting of the project acoustic behaviour; **c** Plotting by pavement with indication of non-attendance area to the minimum level of the standard; **d** Plotting of non-compliant enclosures throughout the building; **e** Aerial driving noise analysis on tinning walls; and **f** Impact noise analysis

The individual modeling results of the elements were used in the other program, Sonarchitect, to simulate the overall situation of acoustic performance of the building. This second modeling allows simulating the field behavior of the project, and the results present the same field indices, in this case the standardized difference of weighted level of the airnoise of horizontal partitions (D_{nTW}) and the level of sound pressure of standard impact ($L'_{nT,W}$), thus allowing confrontation of the measured values with the simulated values.

Finally, the values obtained experimentally and numerically were presented according to the sketch of Table 12. According to the differences found, the simulations of some of the cases in which this difference (simulated versus measured) exceeded 2 dB were redone, with the function of evaluating whether the modeling procedures influenced the assertiveness of the values found in the simulations. These procedures included variation in the adjustments of construction material building for some of the cases, where the damping factor or specific density was corrected.

Table 12 Model used to present and calculate the difference between in-field values and simulated values

		In-field tests	Numerical results SONARCHITECT	
Environment	Criterion (dB) (NBR 15575-4 & NBR 15575-3)	Result (dB) $D_{nT,w}$ (aerial noise) $L'_{nT,w}$ (impact noise)	Result (dB) $D_{nT,w}$ (aerial noise) $L'_{nT,w}$ (impact noise)	Δ (dB) SC-MC
		MC	SC	
Emission: Environment x Reception: Environment y	Minimum: _ to _ Intermediate:_ to _ Superior: ≥ or ≤	NA M I S	NA M I S	SC-MC

where: MC is the value measured in in-field tests; SC is the calculated value by numerical simulation; Δis the SC versus MC difference (Numerical simulation—in-field measurement); NA = Do not meets the criterion; M = Meets at minimum level; I = Meets at intermediate level; S = Meets at superior level

2.3 Results Evaluation Criteria

From the comparison of the simulated values with the measured values, and the calculation of the difference in absolute values of these results, the percentage of values whose difference between the simulation and measurement result was 0 dB, 1 dB, 2 dB, 3 dB and above 3bB, respectively, creating a scale of optimal, good, acceptable, bad and bad, respectively, according to the classification illustrated in Fig. 7.

This "acceptable" classification adopted was based on measurement uncertainty, provided for in NBR 16283-2 (2018), whose calculations made in some samples, resulted in an average value of approximately ±2 dB for such uncertainty.

Thus, taking into account that the measurement value may vary in an acceptable range of ±2 dB, the situation in which there was no difference between the absolute values of the simulation and the measurement was adopted as "optimal" when there was no difference between the absolute values of the simulation and the measurement, meaning that up to 2 dB there is a possibility that the field values are equal to the simulated, because of the uncertainty range of the measurement. With the difference of ±3 dB, the sample leaves the acceptable range, with the simulation distancing itself ±1 dB of the measurement uncertainty range, classifying the sample as "bad", and values higher than (3 dB were identified as "poor", in the sense that there was

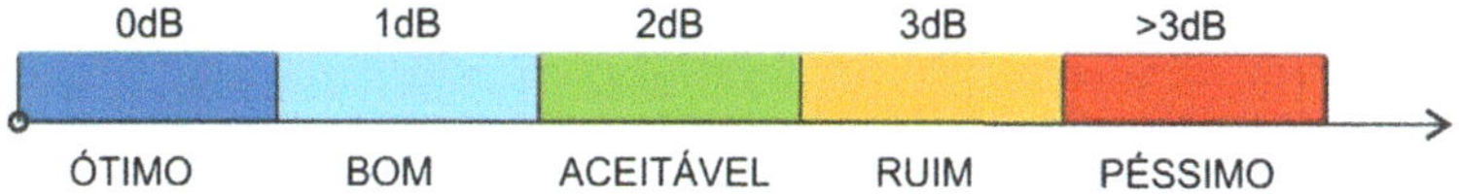

Fig. 7 Evaluation rating adopted

greater incompatibility in the simulation in relation to what would be measured in the field.

After such analyses, we were able to filter those results whose divergences are extreme and analyze such discrepancies on a case-by-case basis, in order to find a homogeneity of the comparison data and evaluate which parameters influenced such results.

3 Description of the Case Studies

In this work, 14 case studies were addressed, all multifamily housing buildings. The construction system adopted for floor systems varied between alveolar/ribbed concrete slabs or solid concrete slabs. Table 13 presents the materials and construction systems of the case studies, in addition to the number of evaluations performed; the acronyms adopted to identify the simulations of this work are identified in Table 14.

Table 13 Simulated case study building systems

Building	SVH	Quantity	
		SVH–RA	SVH-RI
I-1	RC (10) + CP (50) + LJ (50) + AR (60) + FG (20) RC (15) + CP (25) + LJ (50) + AR (100) + FG (20)	8	4
I-2	RC (10) + CP (50) + LJ (50) + AR (100) + FG (20) RC (10) + CP (50) + LJ (50) + AR (200) + FG (20)	10	5
I-3	RC (10) + CP (70) + LJ (60) + AR (260) + FG (20) RC (10) + CP (60) + LJ (60) + AR (260) + FG (20)	7	7
I-4	RC (10) + CP (30) + LJ (60) + AR (110) + FG (20) RC (10) + CP (30) + LJ (60) + AR (200) + FG (20)	4	4
I-5	RC (10) + CP (30) + LJ (60) + AR (100) + FG (30) RC (10) + CP (30) + LJ (60) + AR (100) + FG (20)	3	3
I-6	RC (20) + CP (35) + LJ (60) + AR (200) + FG (20)	20	20
I-7	RC (20) + CP (65) + LJ (60) + AR (120) + FG (20)	10	10
I-8	RC (10) + CPE (40) + CP (30) + LJ (50) + AR (120) + FG (20) RC (10) + CP (70) + LJ (50) + AR (120) + FG (20)	8	8
I-9	RC (20) + CP (40) + LJ (70) + AR (120) + FG (20)	8	8
I-10	RC (10) + CP (25) + LJM (100)	8	8
I-11	RC (10) + CP (30) + LJM (100)	8	8
I-12	RC (10) + CP (5) + LJM (100)	8	8
I-13	RC (15) + CP (30) + LJM (100)	6	6
I-14	RC (10) + CP (5) + LJM (100) RC (15) + LJM (100)	8	3
	Total	**116**	**102**

Table 14 Identification of simulated building materials and systems

Acronym	Material/Systems
RC	Ceramic coating
CP	Counterfloor
LJ	Ribbed concrete slab (capping only)
LJM	Solid concrete slab
CPE	Special floor—isofloc trace
MAF	Asphalt blanket with acab. acrylic resin
AR	Air space

As in this study, the normative criteria of acoustic performance of air noise bedroom of external partition systems (SVVE) were not evaluated, restricting the scope to the verification of the values of the air noise insulation of horizontal partitions (SVH-RA) and impact noise bedroom values (SVH-RI), the characterization of external partition systems and the amount of SVVE measurements was not presented in the summary of Table 13, since such data were not analyzed, however, in the individual descriptions of the case studies, it will be mentioned which systems of the external vertical partitions were used by the respective properties, for better overall characterization of each project and because such data were used as input information in the computer simulations, even though the acoustic performance of such systems (SVVE) was not analyzed.

3.1 *Building I-1*

The building I-1 is located in the city of Recife, composed of 2 residential towers, with 22 floors type per tower, being 4 apartments per floor, with an area of 62.63 m^2 and 74.64 m^2 per apartment. Each housing unit has 2 or 3 bedrooms, being 1 suite, living/dining room, kitchen and service area, as can be seen in the floor plan reproduced in Fig. 8.

Sound insulation tests were performed in apartments of the 5th and 6th floors. The floor system adopted was ribbed slab in reinforced concrete, with a capping thickness of 5.0 cm, plus a 2.5 cm thick floor and ceramic coating. The ceiling has a plaster lining of 2.0 cm thick, 20.0 cm away from the bottom of the rib, which is called air space. The left foot adopted by the design, which is the distance between floors, was 3.00 m. The floor construction system adopted by the building I-1 is:

- Ceramic (1.0 cm plate + 1.0 cm adhesive mortar) + Counterfloor 2.5 cm + Ribbed slab 5.0 cm + plaster lining (air space 20.0 cm).

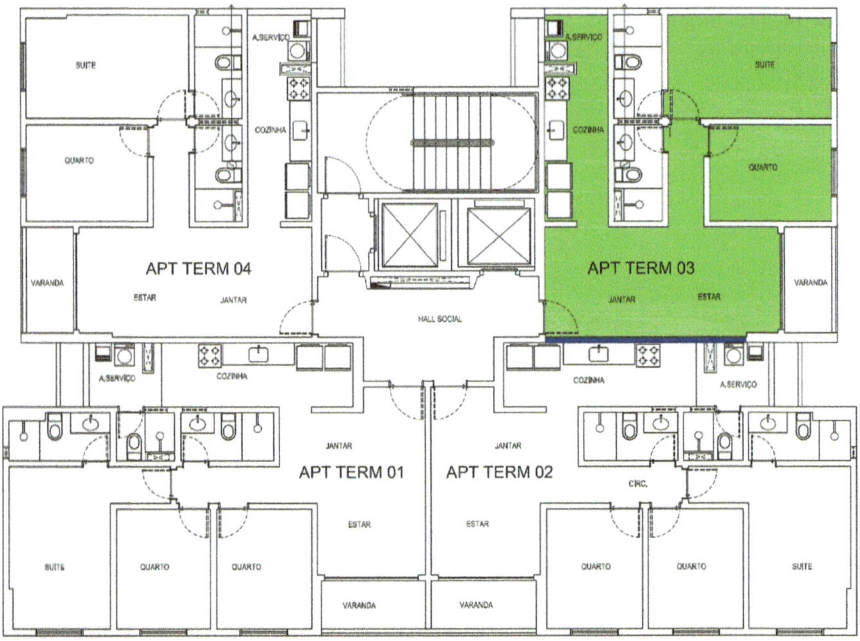

Fig. 8 Plant of building I-1: Detail of the in-field tests

3.2 *Building I-2*

The building I-2 is located in the city of Recife/PE, with 27 type floors, being 4 apartments per floor, with an area of 61.20 m^2 per apartment of terminations 01 and 04, and 61.08 m^2 for apartments of terminals 02 and 03. Each housing unit has 2 bedrooms, 1 suite, living/dining room, kitchen/service area, social bathroom and balcony, as sketched in Fig. 9.

Sound insulation tests were performed in apartments of the 1st, 2nd and 7th floors. The floor system adopted consisted of ribbed slab in reinforced concrete, with a capping of 5.0 cm thickness, plus a 5.0 cm thick cementitious mortar base. The ceiling features a plaster lining of 2.0 cm thick, 20.0 cm away from the bottom of the rib, called air space. Finally, the floor construction system of building I-2 is.

- Counterfloor 5.0 cm + ribbed slab 5.0 cm (capping) + plaster lining (air space 20.0 cm)

3.3 *Building I-3*

The building I-3 is located in the city of Aracaju/SE, composed of 2 towers with 5 type floors per tower. Each type floor has 6 housing units, being 2 units with an area

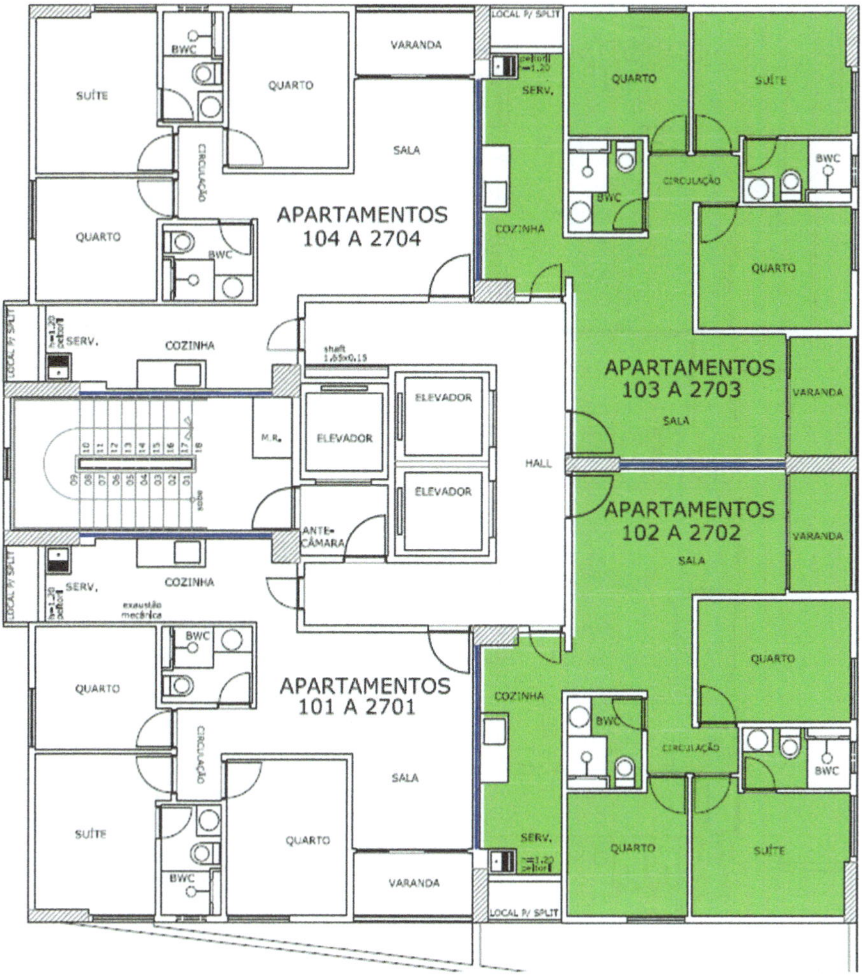

Fig. 9 Plant of building I-2: Detail of the in-field tests

of 80.00 m^2, 2 units with an area of 71.83 m^2 and other 2 units with 79.55 m^2. Each housing unit has 2 and 3 bedrooms, being 1 suite, living/dining room, kitchen/service area, social bathroom and balcony, as can be seen in Fig. 10.

Sound insulation tests were performed in apartments of 5th and 6th floors, and the floor construction system of the partitions adopted by the building I-3 is.

- Ceramic plate 0.5 cm + adhesive mortar 0.5 cm + Back floor 7.0 cm + ribbed slab 6.0 cm (capeamento) + plaster lining (air space 26.0 cm).

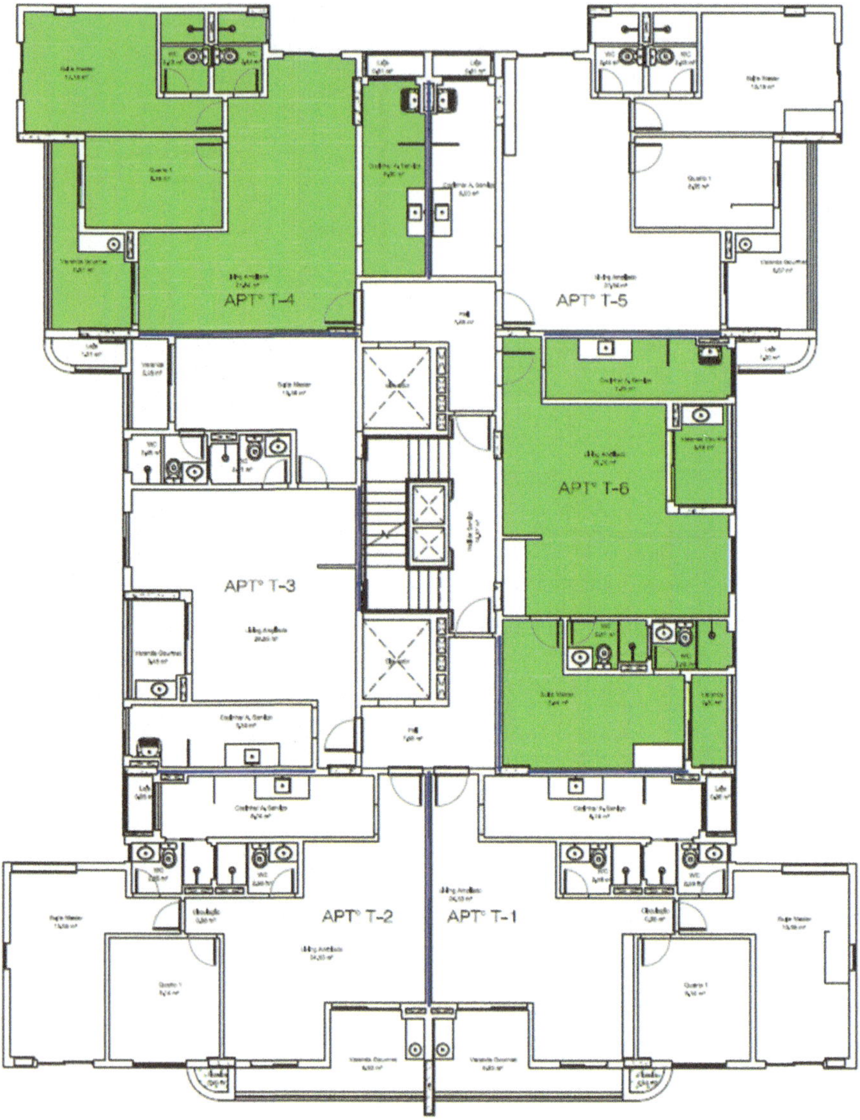

Fig. 10 Plant of building I-3: Detail of the in-field tests

3.4 *Building I-4*

The building I-4 is in the city of João Pessoa/PB, with 18 type floors, being 4 housing units per floor. One of the units of the floor has an area of 75.62 m^2, another has an area of 76.44 m^2 and the two remaining units of the floor has an area of 63.00 m^2.

Each housing unit has 2 and 3 bedrooms, being in both cases 1 suite, living/dining room, kitchen/service area, social bathroom and balcony, as can be seen in Fig. 11.

Sound insulation tests were performed in apartments of the 16th and 17th floors, and the floor construction system of the partitions adopted by the building I-4 is.

- Ceramic plate 0.5 cm + adhesive mortar 0.5 cm + floor 3.0 cm + ribbed slab 6.0 cm (capping) + plaster lining (air space 20.0 cm).

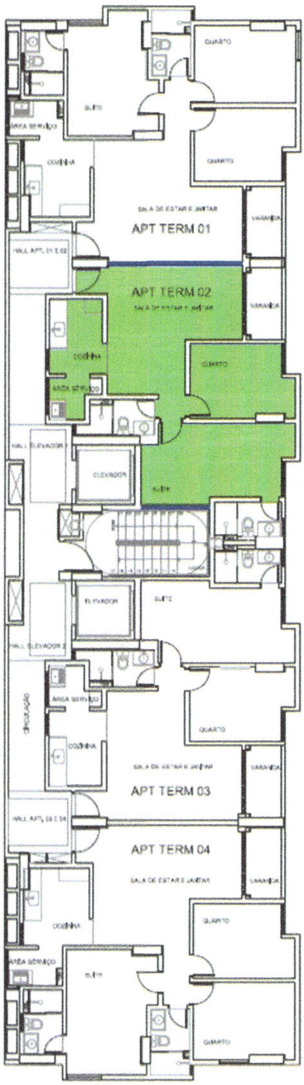

Fig. 11 Plant of building I-4: Detail of the in-field tests

3.5 Building I-5

The building I-5 is located in the city of Recife/PE, composed of 1 tower with 31 floors, 3 apartments per floor, with an area of 49.24 m^2 for apartment termination 01, area of 50.14 m^2 for apartment of termination 02 and area of 50.20 m^2 for the termination apartment 03. Each housing unit has 2 bedrooms, one of them suite, living/dining room, kitchen/service area and social bathroom, as sketched in Fig. 12.

Sound insulation tests were performed in apartments of the 1st and 2nd floors and the floor construction system of the partitions adopted by the building I-5 is.

- Ceramic plate 0.5 cm + adhesive mortar 0.5 cm +backfloor 3.0 cm + ribbed slab 6.0 cm (capeamento) + plaster lining (air space 11.0 cm).

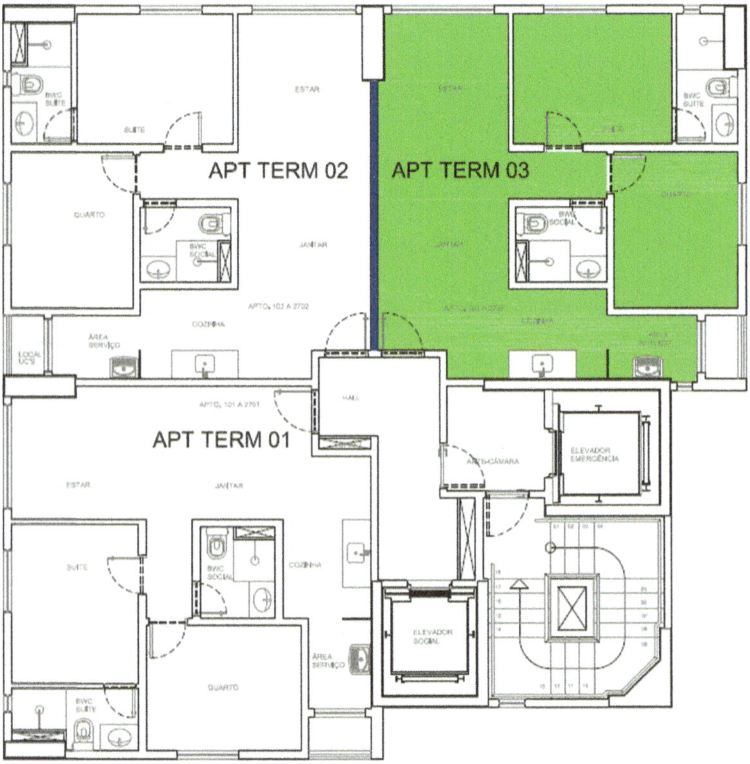

Fig. 12 Plant of building I-5: Detail of the in-field tests

3.6 Building I-6

The building I-6 is located in the city of Recife/PE, composed of 1 tower with 14 floors, being 4 apartments per floor, 2 units with an area of approximately 64.94 m^2 and the other 2 units with an area of 62.54 m^2. Each housing unit has 3 bedrooms, 1 suite, living/dining room, kitchen/service area, social bathroom and balcony, as can be seen in Fig. 13.

Sound insulation tests were performed in apartments of the 1st, 2nd and 3rd floors, and the floor construction system of the building I-6 is.

- Porcelain tile 1.0 cm + adhesive mortar 1.0 cm + floor 3.5 cm + ribbed slab 6.0 cm + plaster lining (air space 10.0 cm).

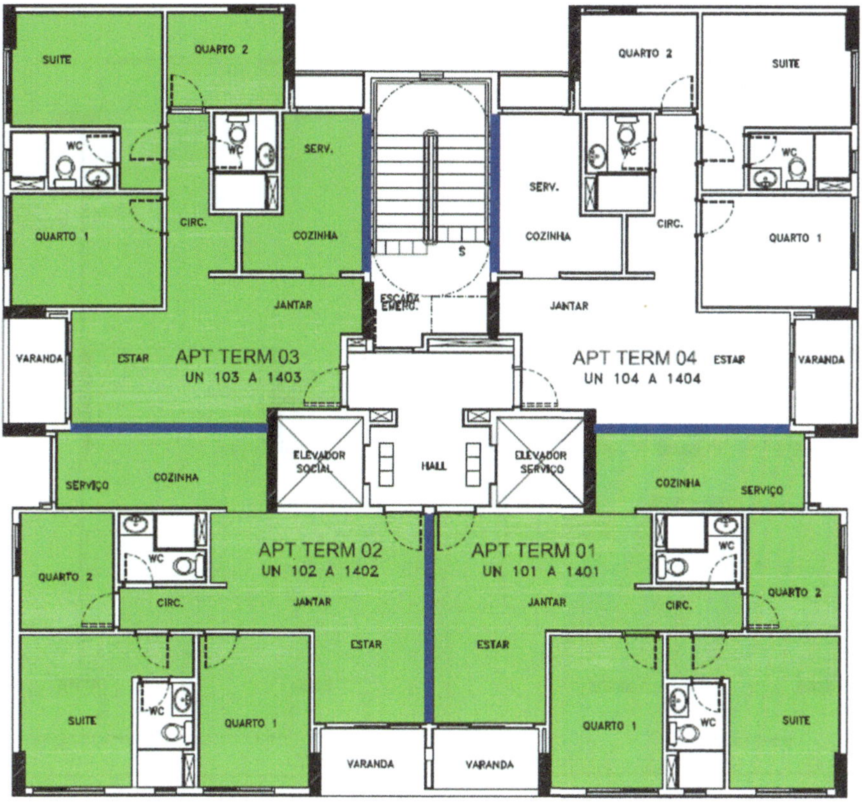

Fig. 13 Plant of building I-6: Detail of the in-field tests

3.7 Building I-7

The building I-7 is located in the city of João Pessoa/PB, composed of 1 tower with 8 floors and 1 duplex floor (penthouse), being 2 housing units per floor, with an area of approximately 61.00 m^2 per apartment. Each housing unit has 3 suites, 2 bathrooms, living/dining room, kitchen, service area and balcony, as can be seen in Fig. 14. The duplex apartments have a penthouse with gourmet area, swimming pool free area and bedroom with bathroom.

Sound insulation tests were performed in apartments of the 1st and 2nd floors, and the floor construction system of the building I-7 is.

- Porcelain tile 1.0 cm + adhesive mortar 1.0 cm + Back floor 6.5 cm + ribbed slab 6.0 cm + plaster lining (air space 12.0 cm).

Fig. 14 Plant of building I-7: Detail of the in-field tests

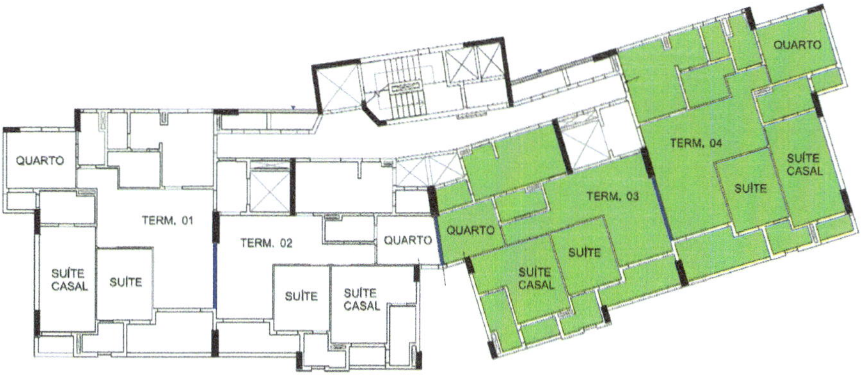

Fig. 15 Plant of building I-8: Detail of the in-field tests

3.8 Building I-8

The building I-8, located in the city of Fortaleza/CE, is a composed of 2 towers, with 17 floors per tower. Each floor has 4 housing units, with an area of approximately 98 m^2 and 118 m^2. The apartments have 3 suites, 1 reversible, living/dining room, kitchen, service area, bedroom and bathroom service and balcony, as can be seen in Fig. 15.

Sound insulation tests were performed in apartments of the 15th and 16th floors of tower 1, and the floor construction system of the building I-8 is.

- Porcelain tile 0.5 cm + adhesive mortar 1.0 cm + floor with ISOFLOC 4.0 cm + floor 3.0 cm + ribbed slab 5.0 cm (capeamento) + plaster lining (air space 12.0 cm).

3.9 Building I-9

The building I-9, located in the city of Cabo de Santo Agostinho/PE, is composed of 3 towers, having the towers 12 floors and 1 floor cover, being 4 apartments per floor, with an area of approximately 127.40 m^2 per housing unit. Each apartment has 4 suites, living/dining room, kitchen/service area, toilet, service bathroom and balcony, as can be seen in Fig. 16.

Sound insulation tests were performed in apartments of the 1st, 2nd and 3rd floors per tower, and the floor construction system of the building I-9 is.

- Ceramic plate 1.0 cm + adhesive mortar 1.0 cm + backfloor 4.0 cm + ribbed slab 7.0 cm (capping) + plaster lining (air space 12.0 cm).

Fig. 16 Plant of building I-9: Detail of the in-field tests

3.10 Building I-10

The building I-10, located in the city of Maceió/AL, consisting of 10 residential towers, each tower with 4 floors, being 1 ground floor and 3 more floors. Each tower has 4 apartments per floor, with an area of approximately 42 m^2 per apartment. The housing unit has 2 bedrooms, star/dining room, kitchen/service area and social bathroom, as can be seen in Fig. 17.

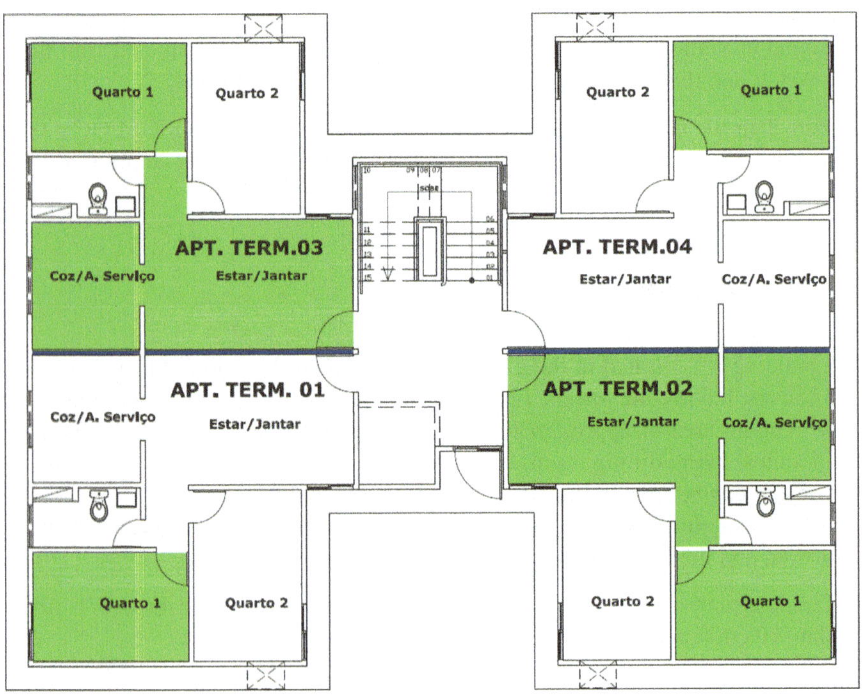

Fig. 17 Plant of building I-10: Detail of the in-field tests

Sound insulation tests were carried out in ground floor and 1st floor apartments. The floor system adopted is solid concrete slab with thickness of 10.0 cm, with a tread of 2.5 cm, mortar of 0.5 cm and ceramic coating of 0.5 cm, plus a plaster coating layer of 0.5 cm under the slab. The floor construction system adopted by the building I-10 is.

- Ceramic plate 0.5 cm + adhesive mortar 0.5 cm + floor 2.5 cm + self-adhesive concrete slab 10.0 cm + plaster coating 0.5 cm.

3.11 Building I-11

The building I-11, located in the city of Camaçari/BA, is composed of 30 towers, being 25 towers with 3 type floors and 1 ground floor and 5 towers with 4 type floors and 1 ground floor. All towers have 4 housing units per floor, with an area of approximately 43 m^2 per apartment. Each housing unit has 2 bedrooms, with the option of suites in some units, living/dining room, kitchen/service area, social bathroom, service room and balcony, as sketched in Fig. 18.

Sound insulation tests were carried out in ground floor and 1st floorand the floor construction system adopted by the building I-11 is.

Fig. 18 Plant of building I-11: Detail of the in-field tests

- Ceramic plate 0.5 cm + adhesive mortar 0.5 cm + floor 3.0 cm + self-adhesive concrete slab 10.0 cm + roughcast + plaster partitioner + texture.

3.12 Building I-12

The building I-12 is located in the city of Camaçari/BA with 18 residence towers, twined every 2 towers, composed of ground floor and 3 type floors. Each tower has 4 housing units per floor, with an area of approximately 43.00 m^2 per apartment. Between towers, it is included in the project that there is a 20 mm expansion joint. Each housing unit has 2 bedrooms, living/dining room, kitchen/service area and social bathroom, as sketched in Fig. 19.

Sound insulation tests were performed in ground floor apartments, 1st, 2nd and 3rd floors, of towers 1, 15, 17 and 18, and the floor construction system adopted by the building I-12 is.

- Horizontal partition: Ceramic plate 1.0 cm + adhesive mortar 0.5 cm + concrete slab 10.0 cm
- Bedrooms: Ceramic plate 1.0 cm + adhesive mortar 0.5 cm + mechanical protection 0.1 cm + asphalt blanket with acrylic resin finish 0.3 cm + concrete slab 10.0 cm.

Fig. 19 Plant of building I-12: Detail of the in-field tests

3.13 Building I-13

The building I-13, located in the municipality of São Lourenco da Mata/PE, is an enterprise with 21 towers of 4 floors, being 1 ground floor plus 3 types, composed of 4 apartments per floor, with an area of 45.35 m^2 per apartment. The towers are twined every 2 or 3 towers, and, between the towers, is included in the 20 mm expansion joint project. Each housing unit has 2 bedrooms, 1 suite, living/dining room, kitchen/service area, social bathroom and balcony/terrace, as sketched in Fig. 20.

Sound insulation tests were performed in apartments of the 1st and 2nd floors, of towers 11 and 12, and the floor construction system adopted by the building I-13 is.

- Ceramic plate 0.5 cm + adhesive mortar 0.5 cm +backfloor 3.0 cm + ribbed slab 6.0 cm (capeamento) + plaster lining (air space 11.0 cm).

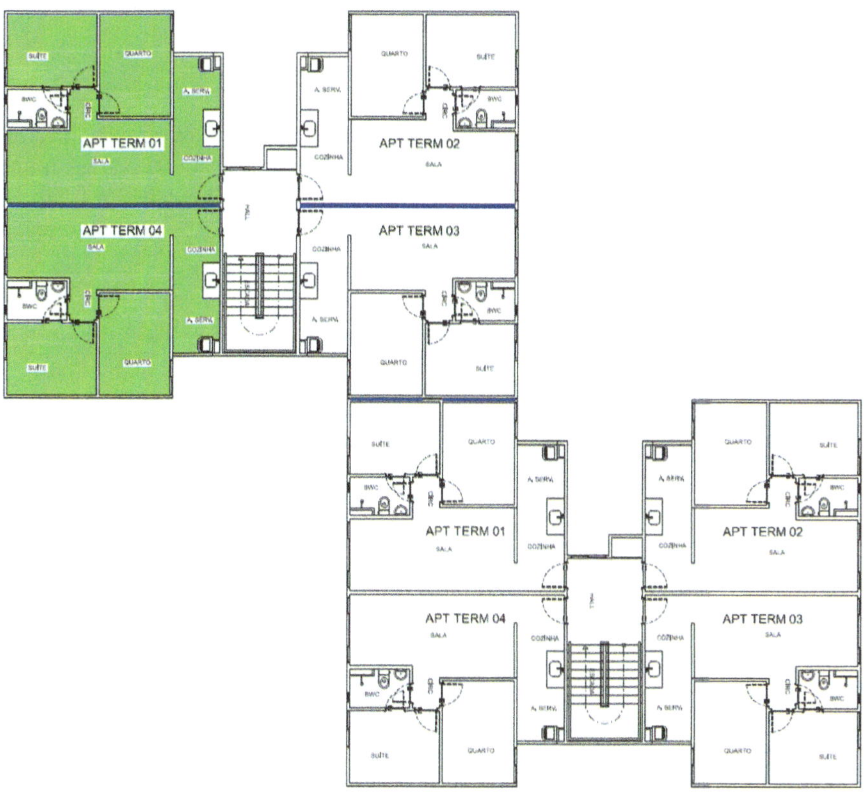

Fig. 20 Plant of building I-13: Detail of the in-field tests

3.14 Building I-14

The buildingI-14, located in the city of Paulista/PE, is a development with 20 towers, with 4 floors, being 1 ground floor and 3 types. Among twin towers, it is included in the project that there is a 20 mm expansion joint. Each tower has 4 apartments per floor, with an area of 37.49 m^2 per apartment. Each housing unit has 2 bedrooms, living/dining room, kitchen/service area and social bathroom, as can be seen in Fig. 21.

Sound insulation tests were performed in apartments of the 1st and 2nd floors of towers 10 and 11, and the floor construction system adopted by the building I-14 is:

- Ceramic plate 1.0 cm + adhesive mortar 0.5 cm + Concrete slab 10.0 cm (Concrete slab 10.0 cm).

4 Results and Discussion

This section presents the results of computational simulations of acoustic performance performed in the 14 case studies and the discussions about the values found in comparison to the values presented in the field trials. SVH evaluations will be presented, divided into air noise—RA and impact noise—RI.

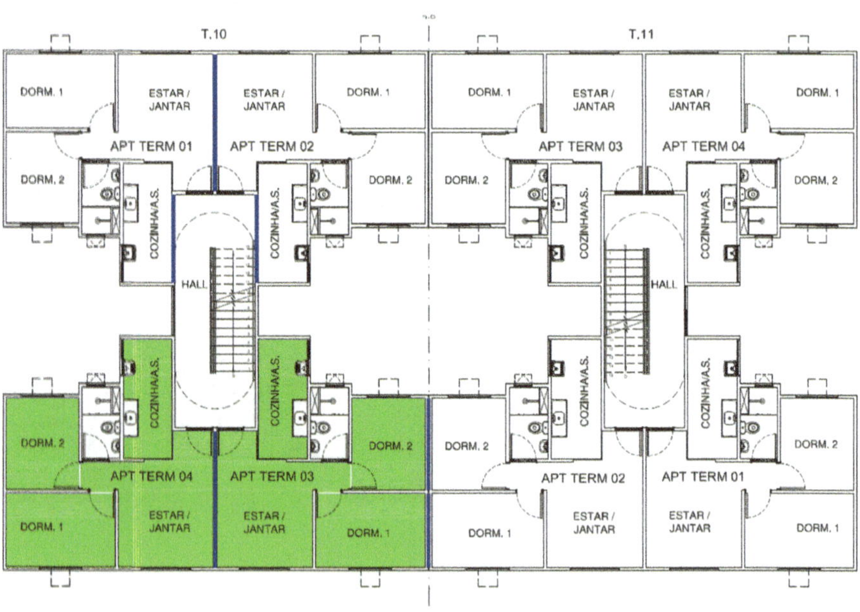

Fig. 21 Plant of building I-14: Detail of the in-field tests

4.1 *Horizontal Partition Systems—SVH*

For horizontal partition systems (SVH), 228 simulations were performed, 116 evaluation studies for air noise bedroom (SVH-RA) and 102 studies on impact noise bedroom (SVH-RI), according to the summary presented in Table 15.

4.1.1 Horizontal Partition Systems—Air Noise Insulation—SVH-RA

Figure 22 presents a graph where the computational simulation values and field measurement values of all 116 SVH-RA assays are plotted, including the respective uncertainty range of ±2 dB for field measurement. The data in this chart are presented in the order of the analysis of the case studies. Figure 23 shows the dispersion of the calculated difference between the values of the computational simulation and the field measurement, but no longer in order of the case studies, but in ascending order of the calculated variation.

Table 15 Number of SVH tests performed

Building	SVH	Quantity	
		SVH-RA	SVH-RI
I-1	RC (10) + CP (50) + LJ (50) + AR (60) + FG (20) RC (15) + CP (25) + LJ (50) + AR (100) + FG (20)	8	4
I-2	RC (10) + CP (50) + LJ (50) + AR (100) +FG (20) RC (10) + CP (50) + LJ (50) + AR (200) +FG (20)	10	5
I-3	RC (10) + CP (70) + LJ (60) + AR (260) + FG (20) RC (10) + CP (60) + LJ (60) + AR (260) + FG (20)	7	7
I-4	RC (10) + CP (30) + LJ (60) + AR (110) + FG (20) RC (10) + CP (30) + LJ (60) + AR (200) + FG (20)	4	4
I-5	RC (10) + CP (30) + LJ (60) + AR (100) + FG (30) RC (10) + CP (30) + LJ (60) + AR (100) + FG (20)	3	3
I-6	RC (20) + CP (35) + LJ (60) + AR (200) + FG (20)	20	20
I-7	RC (20) + CP (65) + LJ (60) + AR (120) + FG (20)	10	10
I-8	RC (10) + CPE (40) + CP (30) + LJ (50) + AR (120) + FG (20) RC (10) + CP (70) + LJ (50) + AR (120) + FG (20)	8	8
I-9	RC (20) + CP (40) + LJ (70) + AR (120) + FG (20)	8	8
I-10	RC (10) + CP (25) + LJM (100)	8	8
I-11	RC (10) + CP (30) + LJM (100)	8	8
I-12	RC (10) + CP (5) + LJM (100)	8	8
I-13	RC (15) + CP (30) + LJM (100)	6	6
I-14	RC (10) + CP (5) + LJM (100) RC (15) + LJM (100)	8	3
	Total	**116**	**102**

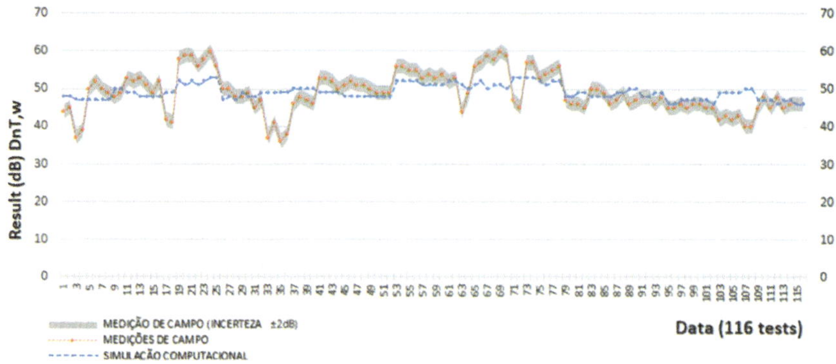

Fig. 22 Scatter plot of results (DnT,w) measured in field and simulated—SVH-RA

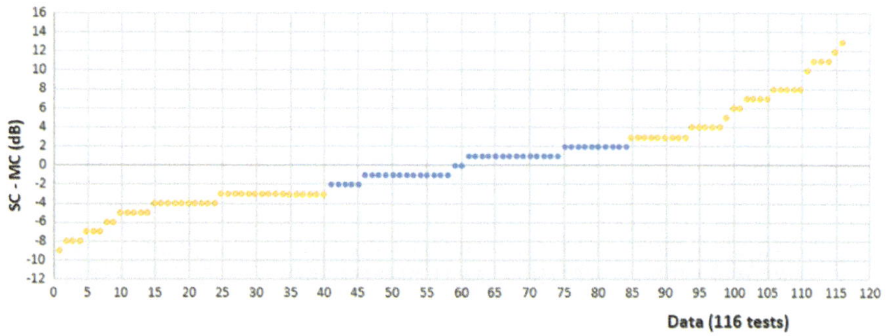

Fig. 23 Dispersion plot of SC-MC (in dB)—SVH-RA

Regarding the calculated values of the computational simulation versus field measurement difference (SC vs. MC), the condition of SC being greater or lower than MC was disregarded, considering the values in module for this difference, and then a histogram of the relative frequencies of this difference SC versus MC was assembled, as shown in Fig. 24.

For the total SVH-RA sample, that is, for the total of 116 trials, 1.72% of the sample presented equal values in SC and MC, being therefore classified as "optimal" result, 23.28% presented a difference of 1 dB, for which it was classified as a "good" result and 12.93% presented a difference of 2 dB, whose classification adopted was "acceptable", totaling 37.93% of the sample within the considered uncertainty range of ±2 dB, this being the acceptable margin considered for the difference between simulation and.

In addition to these percentages, 21.55% presented a difference of 3 dB, classified as a "bad" result and 40.52% presented results from 4 dB, reaching in this sample some values of difference of up to 13 dB between SC and MC, whose classification received was of "poor" level, totaling 62.07% of bad to poor for this sample of SVH-RA.

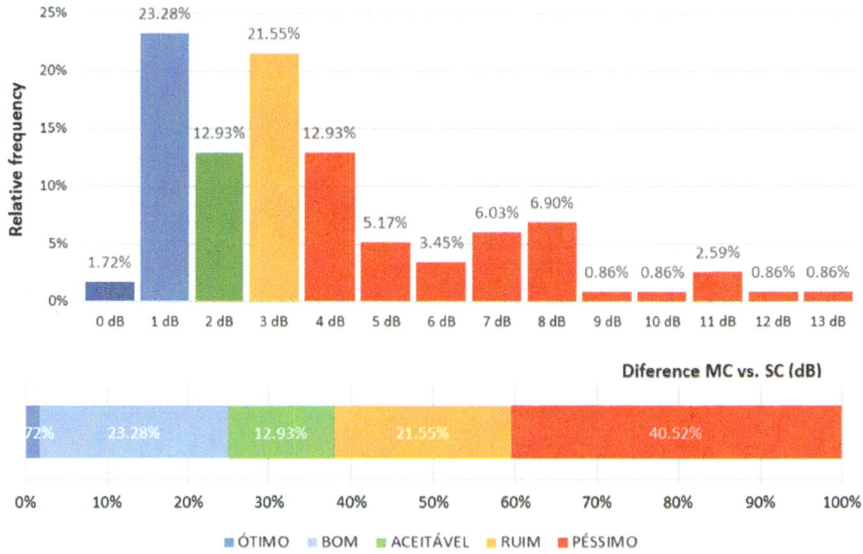

Fig. 24 Histogram of relative frequency of difference SC versus MC for general SVH-RA assays

Another analysis performed with the data of computational simulation and field measurement was the amount of results in which the SC was higher than the MC or vice versa, that is, for this sample it was verified that, of the 116 SVH-RA tests, 48.28% presented computational simulation results higher than the results of field measurement (SC > MC); 50% of the tests presented higher field measurement than the computational simulation result (SC < MC) and 1.72% of the tests presented equal values between field measurement and computational simulation (SC = MC), as illustrated in Fig. 25.

Fig. 25 Comparison of the results of SC versus MC—SVH-RA

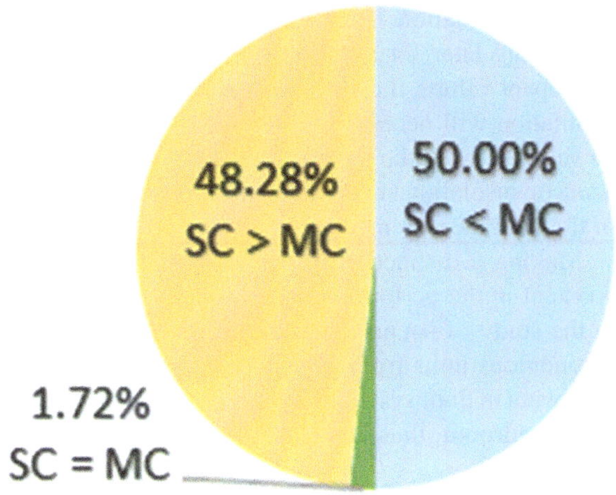

Table 16 Number of tests analyzed for SVH-RA according to element types of the standard NBR 15575-3—Limits for the Engineering Method

Number of Tests performed	Types of SVH-RA Elements and Constant Limits in NBR 15575-3		
	Element	$D_{nT,w}$ (dB)	Performance Level
65	Floor system by separateing autonomous housing units from areas where one of the precincts is a bedroom	**45 to 49**	**M**
		50 to 54	I
		≥55	S
51	Floor system separates autonomous housing units from common areas of eventual traffic, such as corridors and staircase on the pavements, as well as on distinct floors (in autonomous units without one of the enclosures being bedroom)	**40 to 44**	**M**
		45 to 49	I
		≥ 50	S
0	Floor system separates autonomous housing units from common areas for collective use for leisure and sports activities, such as home theater, gyms, ballrooms, games room, collective bathrooms and changing rooms, kitchens and collective laundries	**45 to 49**	**M**
		50 to 54	I
		≥ 55	S

This result, disregarding any uncertainty range, suggests that for SVH-RA, the probability that the computational simulation is greater than the field measurement is equal to the opposite situation, that is, the measurement presents values greater than the simulation.

Starting later for a more detailed analysis, and removing the atypical cases of discrepant values, it can be analyzed if, for SVH-RA, the initial hypothesis that the simulation will necessarily present values greater than the field measurement will be valid, since in the computational simulation the conditions for the calculation of acoustic insulation are predictable and considered ideal, as opposed to the conditions of the field measurement.

Making a distinction in the analysis regarding the type of separation element constant in the performance standard, it was found that of the 116 tests performed in the study of sound insulation in SVH-RA, 65 were in floor systems separating autonomous units from areas in which one of the enclosures was bedroom and 51 tests were in floor systems separating autonomous units without one of the enclosures being bedroom, thus taking into account two of the situations provided for in the standard. Table 16 shows the number of tests performed for each classification of the SVH-RA elements provided for in the performance standard.

Condition I—SVH-RA in floor systems between autonomous housing units in which one of the enclosures is a bedroom Table 17 presents the results for SVH of the 65 field measurement and computational simulation tests for floor systems that separate autonomous housing units in which one of the enclosures is a bedroom, with the respective calculated difference (SC—MC).

Figure 26 presents a graph where the computational simulation values and field measurement values of all 65 SVH-RA tests are plotted for condition I in which there is a bedroom environment, including the respective uncertainty range of ±2 dB for in-field measurement, presented in the order of the trials of each case study as shown in Table 17. Figure 27 shows the dispersion of the calculated difference between the values of the computational simulation and the field measurement, in ascending order of the calculated variation.

Regarding the calculated values of the computational simulation versus field measurement difference (SC vs. MC), the condition of SC being greater or lower than MC was disregarded, considering the module values for this difference, and then a histogram of the relative frequencies of this difference SC versus MC was assembled, as shown in Fig. 28.

For SVH-RA condition I, between floors with bedroom environment, of the total of 65 trials, 3.08% of the sample presented equal values in SC and MC, and this portion of the sample classified as "optimal", 27.69% presented a difference of 1 dB, for which it was classified as a "good" result and 12.31% presented a difference of 2 dB, whose classification was "acceptable", totaling 43.08% of the sample within the considered uncertainty range of ±2 dB, this being the acceptable margin considered for the difference between the simulation and the measurement of the present study.

In addition to these percentages, 24.62% presented a difference of 3 dB, receiving this portion of "bad" result and 32.31% presented results from 4 dB, reaching some values of difference of up to 9 dB between SC and MC, whose classification received was of "poor" level, totaling 56.92% from bad to poor for this sample.

Regarding the comparison of how far the computational simulation data are from the field measurement, for this sample (SVH-RA Condition I), it was verified that 24.62% of the 65 tests presented computational simulation results higher than the results of field measurement (SC > MC); 72.31% of the tests presented higher field measurement than the computational simulation result (SC < MC) and 3.08% of the tests presented equal values between field measurement and computational simulation (SC = MC), as shown in Fig. 29.

This result, disregarding any range of uncertainty, is in line with what was expected as an initial hypothesis that the simulation would necessarily present values higher than field measurement, since in the computational simulation the conditions for the calculation of sound insulation are predictable and considered ideal, as opposed to the conditions of field measurement in which several unpredictable factors, may adversely interfere with the results.

Table 17 SVH-RA tests in floor systems between autonomous housing units, in situations where there is a bedroom environment

Criterion (dB) (NBR 15575-3)	Environment	In-field	Numerical	Δ (dB)
Minimum: 45 to 49 Intermediate: 50 to 54 Superior: ≥55	Systems between floors between housing units in which an environment is a bedroom	Result (dB) $D_{nT,w}$	Result (dB) $D_{nT,w}$	SC-MC
I-01	Floor system—Suite Emiss Ap.2003 > Recept Ap.2103	50	47	−3
	Floor system—Suite Emiss Ap.2103 > Recept Ap.2003	52	47	−5
	Floor system—Bedroom Emiss Ap.2103 > Recept Ap.2003	50	47	−3
	Floor system—Bedroom Emiss Ap.2003 > Recept Ap.2103	49	47	−2
I-02	Floor system—Bedroom 1 Emiss Ap.102 > Recept Ap.202	53	49	−4
	Floor system—Bedroom 1 Emiss Ap.103 > Recept Ap.203	52	49	−3
	Floor system—Bedroom 2 Emiss Ap.102 > Recept Ap.202	53	48	−5
	Floor system—Bedroom 2 Emiss Ap.103 > Recept Ap.203	51	48	−3
	Floor system—Bedroom 3 Emiss Ap.102 > Recept Ap.202	49	48	−1

(continued)

Table 17 (continued)

Criterion (dB) (NBR 15575-3)	Environment	In-field	Numerical	Δ (dB)
Minimum: 45 to 49 Intermediate: 50 to 54 Superior: ≥55	Systems between floors between housing units in which an environment is a bedroom	Result (dB) $D_{nT,w}$	Result (dB) $D_{nT,w}$	SC-MC
	Floor system—Bedroom 3 Emiss Ap.103 > Recept Ap.203	52	48	−4
I-03	Floor system—Bedroom 1 Emiss Ap.504 > Recept Ap.604	58	52	−6
	Floor system—Bedroom 2 Emiss Ap.504 > Recept Ap.604	59	51	−8
	Floor system—Bedroom Emiss Ap.506 > Recept Ap.606	59	52	−7
	Floor system—Suite Emiss Ap.504 > Recept Ap.604	56	51	−5
	Floor system—Suite Emiss Ap.506 > Recept Ap.606	58	52	−6
I-04	Floor system—Bedroom Emiss Ap.1602 > Recept Ap.1702	50	47	−3
	Floor system—Suite Emiss Ap.1602 > Recept Ap.1702	50	48	−2
I-05	Floor system—Bedroom Emiss Ap.203 > Recept Ap.103	49	48	−1
	Floor system—Suite Emiss Ap.203 > Recept Ap.103	45	48	3

(continued)

Table 17 (continued)

Criterion (dB) (NBR 15575-3)	Environment	In-field	Numerical	Δ (dB)
Minimum: 45 to 49 Intermediate: 50 to 54 Superior: ≥55	Systems between floors between housing units in which an environment is a bedroom	Result (dB) $D_{nT,w}$	Result (dB) $D_{nT,w}$	SC-MC
I-06	Floor system—Bedroom 1 Emiss Ap.101 > Recept Ap.201	53	49	−4
	Floor system—Bedroom 1 Emiss Ap.102 > Recept Ap.202	53	49	−4
	Floor system—Bedroom 1 Emiss Ap.103 > Recept Ap.203	52	49	−3
	Floor system—Bedroom 1 Emiss Ap.203 > Recept Ap.303	50	49	−1
	Floor system—Bedroom 2 Emiss Ap.101 > Recept Ap.201	51	48	−3
	Floor system—Bedroom 2 Emiss Ap.102 > Recept Ap.202	52	48	−4
	Floor system—Bedroom 2 Emiss Ap.103 > Recept Ap.203	51	48	−3
	Floor system—Bedroom 2 Emiss Ap.203 > Recept Ap.303	51	48	−3
	Floor system—Suite Emiss Ap.101 > Recept Ap.201	50	48	−2
	Floor system—Suite Emiss Ap.102 > Recept Ap.202	49	48	−1

(continued)

Table 17 (continued)

Criterion (dB) (NBR 15575-3)	Environment	In-field	Numerical	Δ (dB)
Minimum: 45 to 49 Intermediate: 50 to 54 Superior: ≥55	Systems between floors between housing units in which an environment is a bedroom	Result (dB) $D_{nT,w}$	Result (dB) $D_{nT,w}$	SC-MC
	Floor system—Suite Emiss Ap.103 > Recept Ap.203	49	48	−1
	Floor system—Suite Emiss Ap.203 > Recept Ap.303	49	48	−1
I-07	Floor system—Suite 1 Emiss Ap.101 > Recept Ap.201	53	51	−2
	Floor system—Suite 1 Emiss Ap.102 > Recept Ap.202	54	51	−3
	Floor system—Suite 2 Emiss Ap.101 > Recept Ap.201	53	51	−2
	Floor system—Suite 2 Emiss Ap.102 > Recept Ap.202	54	51	−3
	Floor system—Suite 3 Emiss Ap.101 > Recept Ap.201	52	52	0
	Floor system—Suite 3 Emiss Ap.102 > Recept Ap.202	53	52	−1
I-08	Floor system—Bedroom Emiss Ap.1503 > Recept Ap.1603	59	51	−8

(continued)

Table 17 (continued)

Criterion (dB) (NBR 15575-3)	Environment	In-field	Numerical	Δ (dB)
Minimum: 45 to 49 Intermediate: 50 to 54 Superior: ≥55	Systems between floors between housing units in which an environment is a bedroom	Result (dB) $D_{nT,w}$	Result (dB) $D_{nT,w}$	SC-MC
	Floor system—couple suite Emiss Ap.1503 > Recept Ap.1603	58	51	−7
	Floor system—Suite Emiss Ap.1504 > Recept Ap.1604	60	51	−9
	Floor system—couple suite Emiss Ap.1504 > Recept Ap.1604	59	51	−8
I-09	Floor system—couple suite Emiss Ap.202S > Recept Ap.302S	53	52	−1
	Floor system—daughter suite Emiss Ap.202 s > Recept Ap.302 s	54	51	−3
	Floor system—Son suite Emiss Ap.202S > Recept Ap.302S	55	52	−3
	Floor system—Visits suite Emiss Ap.102S > Recept Ap.202S	56	52	−4
I-10	Floor system—Bedroom 1 Emiss Ap.01 > Recept Ap.101	50	49	−1

(continued)

Table 17 (continued)

Criterion (dB) (NBR 15575-3)	Environment	In-field	Numerical	Δ (dB)
Minimum: 45 to 49 Intermediate: 50 to 54 Superior: ≥55	Systems between floors between housing units in which an environment is a bedroom	Result (dB) $D_{nT,w}$	Result (dB) $D_{nT,w}$	SC-MC
	Floor system—Bedroom 1 Emiss Ap.02 > Recept Ap.102	50	49	−1
	Floor system—Bedroom 1 Emiss Ap.03 > Recept Ap.103	49	49	0
	Floor system—Bedroom 1 Emiss Ap.04 > Recept Ap.104	46	49	3
I-11	Floor system—Bedroom 1 Emiss Ap.01 > Recept Ap.101	48	49	1
	Floor system—Bedroom 1 Emiss Ap.02 > Recept Ap.102	48	49	1
	Floor system—Bedroom 2 Emiss Ap.01 > Recept Ap.101	46	49	3
	Floor system—Bedroom 2 Emiss Ap.02 > Recept Ap.102	48	49	1
I-12	Floor system—Bedroom 1 Emiss Ap.103 > Recept Ap.203 TB	46	47	1
	Floor system—Bedroom 1 Emis Ap.204 > Recept Ap.304TA	46	47	1

(continued)

Table 17 (continued)

Criterion (dB) (NBR 15575-3)	Environment	In-field	Numerical	Δ (dB)
Minimum: 45 to 49 Intermediate: 50 to 54 Superior: ≥55	Systems between floors between housing units in which an environment is a bedroom	Result (dB) $D_{nT,w}$	Result (dB) $D_{nT,w}$	SC-MC
	Floor system—Bedroom 2 Emiss Ap.103 > Recept Ap.203 TB	45	47	2
	Floor system—Bedroom 2 Emis Ap.204 > Recept Ap.304TA	45	46	1
I-13	Floor system—Bedroom Emiss Ap.101 > Recept Ap.201	42	50	8
	Floor system—Bedroom Emiss Ap.104 > Recept Ap.204	43	50	7
	Floor system—Suite Emiss Ap.101 > Recept Ap.201	42	49	7
	Floor system—Suite Emiss Ap.104 > Recept Ap.204	43	49	6
I-14	Floor system—Bedroom 1 Emiss Ap.103 > Recept Ap.203	45	47	2
	Floor system—Bedroom 1 Emiss Ap.104 > Recept Ap.204	48	47	−1
	Floor system—Bedroom 2 Emiss Ap.103 > Recept Ap.203	45	47	2
	Floor system—Bedroom 2 Emiss Ap.104 > Recept Ap.204	48	47	−1

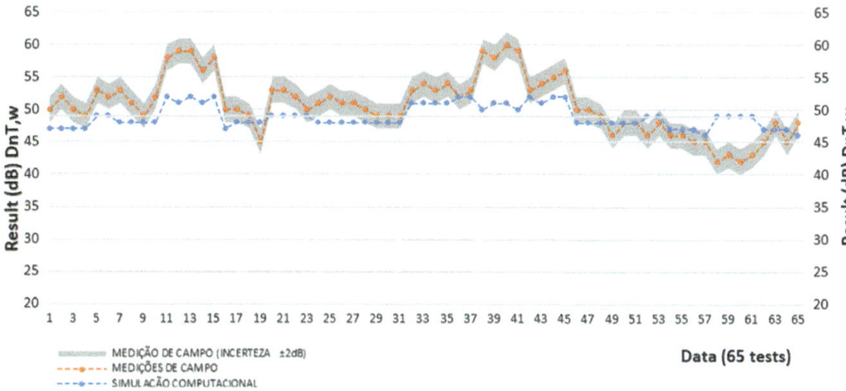

Fig. 26 Scatter plot of results ($D_{nT,w}$) measured in field and simulated—SVH-RA Condition I

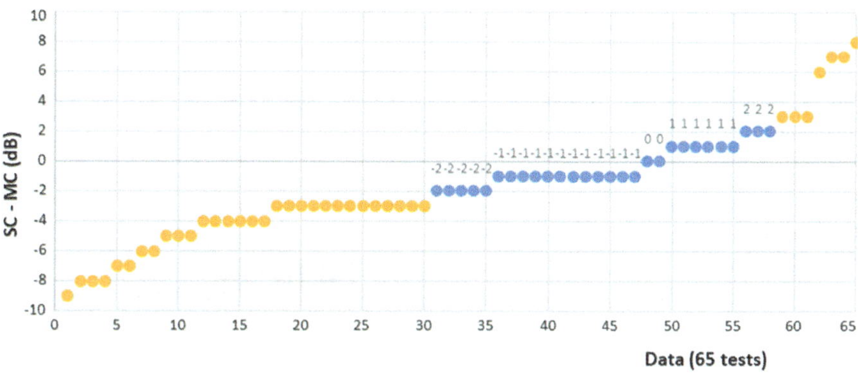

Fig. 27 Dispersion plot of SC-MC (in dB)—SVH-RA Condition I

Condition II—SVH-RA in floor systems between autonomous housing units in which one of the enclosures is not a bedroom Table 18 presents the results for SVH of the 51 in-field measurement and computational simulation tests for floor systems that separate autonomous housing units in which one of the enclosures is not bedroom, with the respective calculated difference (SC-MC).

Figure 30 presents a graph where the computational simulation values and field measurement values of all 51 SVH-RA tests are plotted for condition II in which there is no bedroom environment, including the respective uncertainty range of ±2 dB for field measurement, ordered as shown in Table 18.

Figure 31 shows the dispersion, in ascending order, of the calculated difference between the values of the computational simulation and the field measurement.

Regarding the calculated values of the computational simulation versus field measurement difference (SC vs. MC) presented in Fig. 31, the condition of SC being greater or lower than MC was disregarded, and with the values in module a

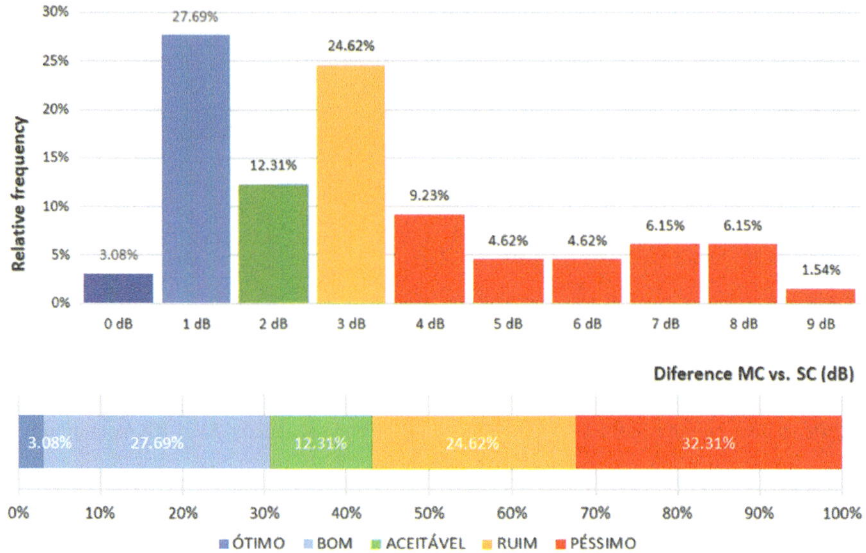

Fig. 28 Histogram of relative frequency of the difference between SC and MC, for SVH-RA Condition I

Fig. 29 Comparison of the results of SC versus MC—SVH-RA Condition I

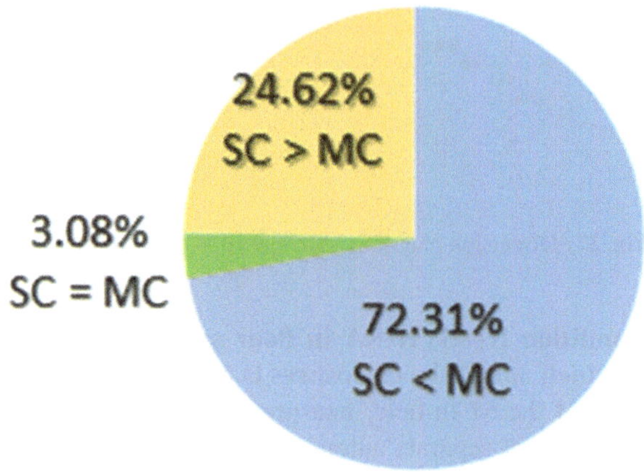

histogram of the relative frequencies of this difference SC versus MC, presented in Fig. 32.

For SVH-RA condition II, that is, for the total of 51 tests of the systems between floors in environments without bedroom, 17.65% of the total tests presented a difference of 1 dB between SC and MC, for which it was classified as "good" result and 13.73% presented a difference of 2 dB, whose classification was "acceptable", totaling 31.37% of the sample within the uncertainty range considered of ±2 dB,

Table 18 SVH-RA tests in floor systems between autonomous housing units, in situations where there is no a bedroom environment

Criterion (dB) (NBR 15575-3)	Environment	In-field	Numerical	Δ (dB)
Minimum: 40 to 44 Intermediate: 45 to 49 Superior: ≥50	Systems between floors between housing units in which an environment is not a bedrrom	Result (dB) $D_{nT,w}$	Result (dB) $D_{nT,w}$	SC—MC
I-01	Floor system—Living room Emiss Ap.2003 > Recept Ap.2103	44	48	4
	Floor system—Living room Emiss Ap.2103 > Recept Ap.2003	45	48	3
	Floor system—Kitchen Emiss Ap.2003 > Recept Ap.2103	37	47	10
	Floor system—Kitchen Emiss Ap.2103 > Recept Ap.2003	39	47	8
I-02	Floor system—Living room Emiss Ap.102 > Recept Ap.202	48	50	2
	Floor system—Living room Emiss Ap.103 > Recept Ap.203	49	50	1
	Floor system—Kitchen Emiss Ap.102 > Recept Ap.202	42	49	7
	Floor system—Kitchen Emiss Ap.103 > Recept Ap.203	41	49	8

(continued)

Table 18 (continued)

Criterion (dB) (NBR 15575-3)	Environment	In-field	Numerical	Δ (dB)
Minimum: 40 to 44 Intermediate: 45 to 49 Superior: ≥50	Systems between floors between housing units in which an environment is not a bedrrom	Result (dB) $D_{nT,w}$	Result (dB) $D_{nT,w}$	SC—MC
I-03	Floor system—Living room Emiss Ap.504 > Recept Ap.604	60	53	−7
	Floor system—Living room Emiss Ap.506 > Recept Ap.606	56	53	−3
I-04	Floor system—Kitchen Emiss Ap.1602 > Recept Ap.1702	48	47	−1
	Floor system—Living room Emiss Ap.1602 > Recept Ap.1702	48	49	1
I-05	Floor system—Living room Emiss Ap.203 > Recept Ap.103	47	49	2
I-06	Floor system—Kitchen Emiss Ap.101 > Recept Ap.201	37	49	12
	Floor system—Kitchen Emiss Ap.102 > Recept Ap.202	41	49	8
	Floor system—Kitchen Emiss Ap.103 > Recept Ap.203	36	49	13

(continued)

Table 18 (continued)

Criterion (dB) (NBR 15575-3)	Environment	In-field	Numerical	Δ (dB)
Minimum: 40 to 44 Intermediate: 45 to 49 Superior: ≥50	Systems between floors between housing units in which an environment is not a bedrrom	Result (dB) $D_{nT,w}$	Result (dB) $D_{nT,w}$	SC—MC
	Floor system—Kitchen Emiss Ap.203 > Recept Ap.303	38	49	11
	Floor system—Living room Emiss Ap.101 > Recept Ap.201	46	50	4
	Floor system—Living room Emiss Ap.102 > Recept Ap.202	48	50	2
	Floor system—Living room Emiss Ap.103 > Recept Ap.203	47	50	3
	Floor system—Living room Emiss Ap.203 > Recept Ap.303	46	50	4
I-07	Floor system—Kitchen Emiss Ap.101 > Recept Ap.201	56	52	−4
	Floor system—Kitchen Emiss Ap.102 > Recept Ap.202	56	52	−4
	Floor system—Living room Emiss Ap.101 > Recept Ap.201	55	52	−3

(continued)

Table 18 (continued)

Criterion (dB) (NBR 15575-3)	Environment	In-field	Numerical	Δ (dB)
Minimum: 40 to 44 Intermediate: 45 to 49 Superior: ≥50	Systems between floors between housing units in which an environment is not a bedrrom	Result (dB) $D_{nT,w}$	Result (dB) $D_{nT,w}$	SC—MC
	Floor system—Living room Emiss Ap.102 > Recept Ap.202	55	52	−3
I-08	Floor system—Kitchen Emiss Ap.1503 > Recept Ap.1603	44	51	7
	Floor system—Kitchen Emiss Ap.1504 > Recept Ap.1604	49	51	2
	Floor system—Living room Emiss Ap.1503 > Recept Ap.1603	56	51	−5
	Floor system—Living room Emiss Ap.1504 > Recept Ap.1604	57	52	−5
I-09	Floor system—Kitchen Emiss Ap.102S > Recept Ap.202S	47	53	6
	Floor system—Kitchen Emiss Ap.202S > Recept Ap.302S	45	53	8
	Floor system—Living room Emiss Ap.102S > Recept Ap.202S	57	53	−4

(continued)

Table 18 (continued)

Criterion (dB) (NBR 15575-3)	Environment	In-field	Numerical	Δ (dB)
Minimum: 40 to 44 Intermediate: 45 to 49 Superior: ≥50	Systems between floors between housing units in which an environment is not a bedrrom	Result (dB) $D_{nT,w}$	Result (dB) $D_{nT,w}$	SC—MC
	Floor system—Living room Emiss Ap.202S > Recept Ap.302S	57	53	−4
I-10	Floor system—Kitchen Emiss Ap.02 > Recept Ap.102	47	49	2
	Floor system—Kitchen Emiss Ap.03 > Recept Ap.103	46	49	3
	Floor system—Living room Emiss Ap.02 > Recept Ap.102	46	50	4
	Floor system—Living room Emiss Ap.03 > Recept Ap.103	45	50	5
I-11	Floor system—Kitchen Emiss Ap.01 > Recept Ap.101	47	50	3
	Floor system—Kitchen Emiss Ap.02 > Recept Ap.102	49	50	1
	Floor system—Living room Emiss Ap.01 > Recept Ap.101	46	50	4

(continued)

Table 18 (continued)

Criterion (dB) (NBR 15575-3)	Environment	In-field	Numerical	Δ (dB)
Minimum: 40 to 44 Intermediate: 45 to 49 Superior: ≥50	Systems between floors between housing units in which an environment is not a bedrrom	Result (dB) $D_{nT,w}$	Result (dB) $D_{nT,w}$	SC—MC
	Floor system—Living room Emiss Ap.02 > Recept Ap.102	47	50	3
I-12	Floor system—Kitchen Emiss Ap.103 > Recept Ap.203 TB	45	46	1
	Floor system—Kitchen Emiss Ap.204 > Recept Ap.304 TA	45	46	1
	Floor system—Living room Emiss Ap.103 > Recept Ap.203 TB	46	47	1
	Floor system—Living room Emiss Ap.204 > Recept Ap.304 TA	45	47	2
I-13	Floor system—Living room Emiss Ap.101 > Recept Ap.201	40	51	11
	Floor system—Living room Emiss Ap.104 > Recept Ap.204	40	51	11
I-14	Floor system—Living room Emiss Ap.103 > Recept Ap.203	45	48	3

(continued)

Table 18 (continued)

Criterion (dB) (NBR 15575-3)	Environment	In-field	Numerical	Δ (dB)
Minimum: 40 to 44 Intermediate: 45 to 49 Superior: ≥50	Systems between floors between housing units in which an environment is not a bedrrom	Result (dB) $D_{nT,w}$	Result (dB) $D_{nT,w}$	SC—MC
	Floor system—Living room Emiss Ap.104 > Recept Ap.204	46	48	2
	Floor system—Kitchen Emiss Ap.103 > Recept Ap.203	46	47	1
	Floor system—Kitchen Emiss Ap.104 > Recept Ap.204	46	47	1

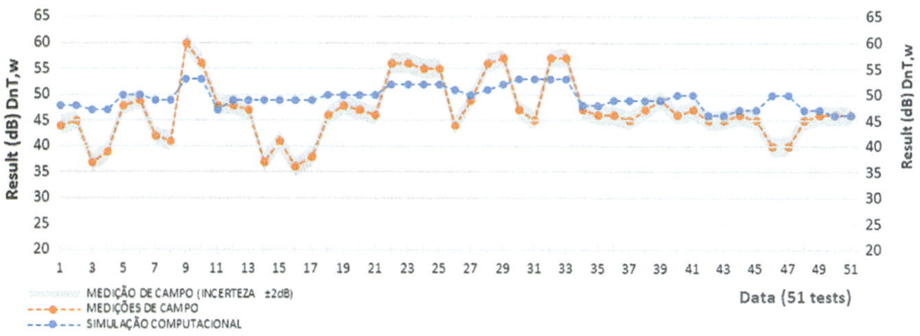

Fig. 30 Scatter plot of results (DnT,w) measured in field and simulated—SVH-RA Condition II

this being the acceptable margin considered for the difference between simulation and.

Apart from these, 17.65% of this sample presented a difference of 3 dB, receiving a "bad" result rating and 50.98% presented results from 4 dB, reaching in this sample the difference values of the order of 13 dB between SC and MC, whose classification received was of "poor" level, totaling 68.63% from bad to poor for this sample.

Regarding the distance between SC and MC, for the SVH-RA sample in Condition II, it was found that 78.43% of the 51 tests presented computational simulation values higher than the field measurement (SC > MC) and 21.57% presented lower field measurement values than computational simulation (SC < MC), as shown in Fig. 33.

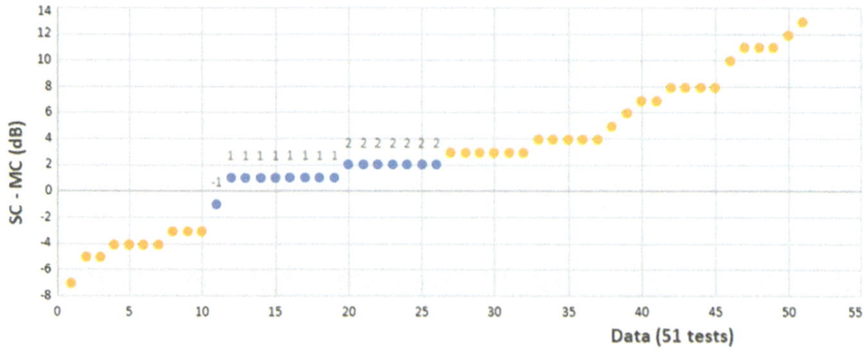

Fig. 31 Dispersion plot of SC-MC (in dB): SVH-RA Condition II

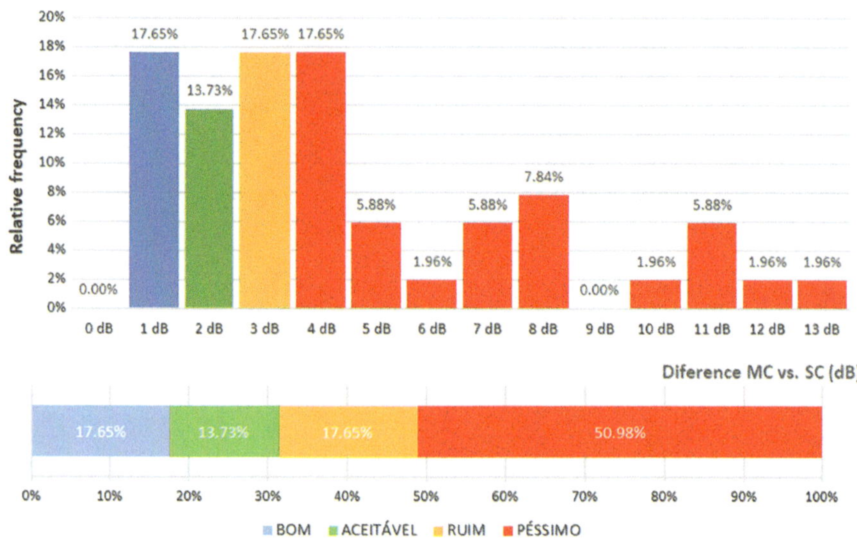

Fig. 32 Histogram of relative frequency of difference SC versus MC for SVH-RA Condition II

For SVH-RA in condition II, this comparative result of the SC and MC indices, disregarding any range of uncertainty, confirms what was expected as an initial hypothesis that the simulation would produce values greater than the field measurement, due to the ideal conditions in the simulation, to the detriment of the observed and possible interferences in the field situation.

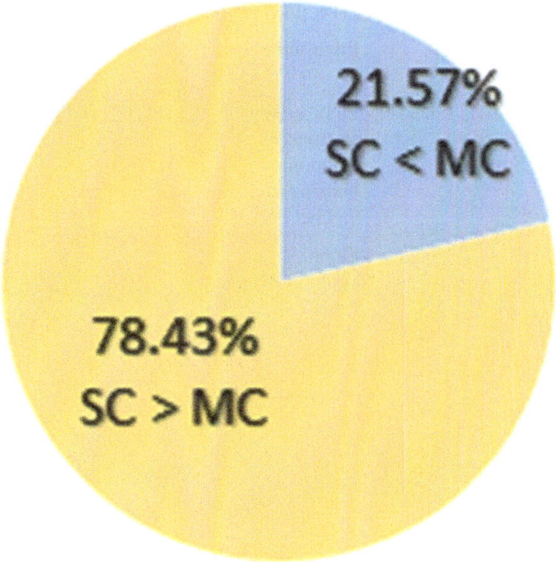

Fig. 33 Comparison of results SC versus MC—SVH-RA Condition II

4.1.2 Horizontal Partition Systems—Impact Noise Insulation—SVH-RI

Table 19 presents the results for SVH-RI of field measurement and computational simulation tests in floor systems by separating autonomous units on different floors, with the respective calculated difference (SC-MC), with a total of 102 samples analyzed.

Figure 34 presents a graph where the computational simulation values and field measurement values of all 102 SVH-RI tests are plotted, including the respective uncertainty range of ±2 dB for field measurement, presented in the order of case studies according to Table 19.

Figure 35 shows the dispersion, in ascending order, of the calculated difference between computational simulation values and field measurement, arranged in this way to facilitate the understanding of the SC versus MC proximity of the data mass.

Regarding the calculated values of the computational simulation versus field measurement difference (SC vs. MC) presented in Fig. 35, the condition of SC being greater or lower than MC was disregarded, considering the module values for this difference, and then followed by assembling a histogram of the relative frequencies of this Difference SC versus MC, as shown in Fig. 36.

For the total SVH-RI sample, that is, for the total of 102 trials, 9.80% of the sample presented equal values in CS and MC, and this portion was classified as "optimal", 18.63% presented a difference of 1 dB, for which it was classified as a "good" result and 17.65% presented a difference of 2 dB, whose classification adopted was "acceptable", totaling 46.08% of the sample within the uncertainty range considered

Table 19 SVH-RI trials on floor systems between autonomous housing units

Criterion (dB) (NBR 15575-3)	Environment	In-field	Numerical	Δ (dB)
Minimum: 66 to 80 Intermediate: 56 to 65 Superior: ≤55	Floor systems by separateing autonomous housing units into distinct floors	Result (dB) L'nT,w	Result (dB) L'nT,w	SC—MC
I-01	Floor system—Kitchen Emiss Ap.2103 > Recept Ap.2003	78	80	2
	Floor system—Bedroom Emiss Ap.2103 > Recept Ap.2003	78	80	2
	Floor system—Living room Emis Ap.2103 > Recep Ap.2003	77	76	−1
	Floor system—Suite Emiss Ap.2103 > Recept Ap.2003	78	78	0
I-02 (AR 200)	Floor system—Kitchen Emiss Ap.102 > Recept Ap.202	74	74	0
	Floor system—Living room Emiss Ap.102 > Recept Ap.202	71	70	−1
	Floor system—Bedroom 1 Emiss Ap.102 > Recept Ap.202	70	75	5
	Floor system—Bedroom 2 Emiss Ap.102 > Recept Ap.202	72	76	4
	Floor system—Bedroom 3 Emiss Ap.102 > Recept Ap.202	74	75	1

(continued)

Table 19 (continued)

Criterion (dB) (NBR 15575-3)	Environment	In-field	Numerical	Δ (dB)
Minimum: 66 to 80 Intermediate: 56 to 65 Superior: ≤55	Floor systems by separateing autonomous housing units into distinct floors	Result (dB) L'nT,w	Result (dB) L'nT,w	SC—MC
I-03	Floor system—Bedroom 1 Emiss Ap.604 > Recept Ap.504	62	72	10
	Floor system—Bedroom 1 Emiss Ap.606 > Recept Ap.506	60	71	11
	Floor system—Bedroom 2 Emiss Ap.604 > Recept Ap.504	62	72	10
	Floor system—Suite Emiss Ap.604 > Recept Ap.504	64	70	6
	Floor system—Suite Emiss Ap.606 > Recept Ap.506	63	69	6
	Floor system—Living room Emiss Ap.604 > Recept Ap.504	58	65	7
	Floor system—Living room Emiss Ap.606 > Recept Ap.506	62	65	3
I-04 (AR 200)	Floor system—Bedroom Emiss Ap.1702 > Recep Ap.1602	80	76	−4

(continued)

Table 19 (continued)

Criterion (dB) (NBR 15575-3)	Environment	In-field	Numerical	Δ (dB)
Minimum: 66 to 80 Intermediate: 56 to 65 Superior: ≤55	Floor systems by separateing autonomous housing units into distinct floors	Result (dB) L'nT,w	Result (dB) L'nT,w	SC—MC
	Floor system—Suite Emiss Ap.1702 > Recept Ap.1602	83	74	−9
	Floor system—Kitchen Emiss Ap.1702 > Recept Ap.1602	78	79	1
	Floor system—Living Bedroom Emiss Ap.1702 > Recept Ap.1602	77	72	−5
I-05	Floor system—Bedroom Emiss Ap.203 > Recept Ap.103	73	81	8
	Floor system—Suite Emiss Ap.203 > Recept Ap.103	75	81	6
	Floor system—Living room Emiss Ap.203 > Recept Ap.103	75	76	1
I-06	Floor system—Kitchen Emiss Ap.201 > Recept Ap.101	70	77	7
	Floor system—Kitchen Emiss Ap.202 > Recept Ap.102	68	77	9

(continued)

Table 19 (continued)

Criterion (dB) (NBR 15575-3)	Environment	In-field	Numerical	Δ (dB)
Minimum: 66 to 80 Intermediate: 56 to 65 Superior: ≤55	Floor systems by separateing autonomous housing units into distinct floors	Result (dB) L'nT,w	Result (dB) L'nT,w	SC—MC
	Floor system—Kitchen Emiss Ap.203 > Recept Ap.103	65	77	12
	Floor system—Kitchen Emiss Ap.303 > Recept Ap.203	72	77	5
	Floor system—Living room Emiss Ap.201 > Recept Ap.101	72	72	0
	Floor system—Living room Emiss Ap.202 > Recept Ap.102	71	72	1
	Floor system—Living room Emiss Ap.203 > Recept Ap.103	71	71	0
	Floor system—Living room Emiss Ap.303 > Recept Ap.203	72	71	−1
	Floor system—Bedroom 1 Emiss Ap.201 > Recept Ap.101	74	76	2

(continued)

of ±2 dB, this being the acceptable margin considered for the difference between simulation and.

In addition to these percentages, 10.78% presented a difference of 3 dB, considered as a "bad" result and 43.14% presented results from 4 dB, reaching values of

Table 19 (continued)

Criterion (dB) (NBR 15575-3)	Environment	In-field	Numerical	Δ (dB)
Minimum: 66 to 80 Intermediate: 56 to 65 Superior: ≤55	Floor systems by separateing autonomous housing units into distinct floors	Result (dB) L'nT,w	Result (dB) L'nT,w	SC—MC
	Floor system—Bedroom 1 Emiss Ap.202 > Recept Ap.102	76	76	0
	Floor system—Bedroom 1 Emiss Ap.203 > Recept Ap.103	63	76	13
	Floor system—Bedroom 1 Emiss Ap.303 > Recept Ap.203	74	76	2
	Floor system—Bedroom 2 Emiss Ap.201 > Recept Ap.101	75	79	4
	Floor system—Bedroom 2 Emiss Ap.202 > Recept Ap.102	72	79	7
	Floor system—Bedroom 2 Emiss Ap.203 > Recept Ap.103	74	79	5
	Floor system—Bedroom 2 Emiss Ap.303 > Recept Ap.203	80	79	−1

(continued)

Table 19 (continued)

Criterion (dB) (NBR 15575-3)	Environment	In-field	Numerical	Δ (dB)
Minimum: 66 to 80 Intermediate: 56 to 65 Superior: ≤55	Floor systems by separateing autonomous housing units into distinct floors	Result (dB) L'nT,w	Result (dB) L'nT,w	SC—MC
	Floor system—Suite Emiss Ap.201 > Recept Ap.101	77	76	−1
	Floor system—Suite Emis Ap.202 > Recept Ap.102	76	76	0
	Floor system—Suite Emis Ap.203 > Recept Ap.103	77	76	−1
	Floor system—Suite Emiss Ap.303 > Recept Ap.203	79	76	−3
I-07	Floor system—Kitchen Emis Ap.201 > Recept Ap.101	69	71	2
	Floor system—Kitchen Emiss Ap.202 > Recept Ap.102	69	71	2
	Floor system—Living room Emiss Ap.201 > Recept Ap.101	72	68	−4
	Floor system—Living room Emiss Ap.202 > Recept Ap.102	72	68	−4

(continued)

Table 19 (continued)

Criterion (dB) (NBR 15575-3)	Environment	In-field	Numerical	Δ (dB)
Minimum: 66 to 80 Intermediate: 56 to 65 Superior: ≤55	Floor systems by separateing autonomous housing units into distinct floors	Result (dB) L'nT,w	Result (dB) L'nT,w	SC—MC
	Floor system—Suite 1 Emiss Ap.201 > Recept Ap.101	73	73	0
	Floor system—Suite 1 Emiss Ap.202 > Recept Ap.102	69	73	4
	Floor system—Suite 2 Emiss Ap.201 > Recept Ap.101	74	74	0
	Floor system—Suite 2 Emiss Ap.202 > Recept Ap.102	72	74	2
	Floor system—Suite 3 Emiss Ap.201 > Recept Ap.101	75	70	−5
	Floor system—Suite 3 Emiss Ap.202 > Recept Ap.102	73	70	−3
I-08 (CPE)	Floor system—Kitchen Emiss Ap.1603 > Recept Ap.1503	66	70	4
	Floor system—Kitchen Emiss Ap.1604 > Recept Ap.1504	70	71	1

(continued)

difference of up to 13 dB between SC and MC, whose classification received was "poor", totaling 53.92% from bad to bad for the SVH-RI sample.

Regarding the distance between SC and MC, for the SVH-RI sample, it was found that 70.59% of the 102 tests presented computational simulation values higher than the field measurement (SC > MC); 21.57% presented lower field measurement values

Table 19 (continued)

Criterion (dB) (NBR 15575-3)	Environment	In-field	Numerical	Δ (dB)
Minimum: 66 to 80 Intermediate: 56 to 65 Superior: ≤55	Floor systems by separateing autonomous housing units into distinct floors	Result (dB) L'nT,w	Result (dB) L'nT,w	SC—MC
	Floor system—Living room Emiss Ap.1603 > Recept Ap.1503	68	67	−1
	Floor system—Living room Emiss Ap.1604 > Recept Ap.1504	67	66	−1
	Floor system—Bedroom Emiss Ap.1603 > Recept Ap.1503	69	72	3
	Floor system—Suite casal Emiss Ap.1603 > Recept Ap.1503	64	69	5
	Floor system—Suite Emiss Ap.1604 > Recept Ap.1504	61	71	10
	Floor system—Suite casal Emiss Ap.1604 > Recept Ap.1504	67	69	2
I-09	Floor system—Kitchen Emis Ap.202S > Recept Ap.102S	66	70	4

(continued)

than computational simulation (SC < MC) and 9.80% of the tests presented equal values between field measurement and computational simulation (SC = MC), as illustrated in Fig. 37.

For the portion of the study referring to the SVH-RI sample, this comparative result of the SC and MC indices, disregarding any range of uncertainty, is in line with what was expected as an initial hypothesis that the simulation would necessarily

Table 19 (continued)

Criterion (dB) (NBR 15575-3)	Environment	In-field	Numerical	Δ (dB)
Minimum: 66 to 80 Intermediate: 56 to 65 Superior: ≤55	Floor systems by separateing autonomous housing units into distinct floors	Result (dB) L'nT,w	Result (dB) L'nT,w	SC—MC
	Floor system—Kitchen Emis Ap.302S > Recept Ap.202S	70	71	1
	Floor system—Living room Emis Ap.202S > Recept Ap.102S	68	67	−1
	Floor system—Living room Emis Ap.302S > Recept Ap.202S	67	66	−1
	Floor system—Suite casal Emis Ap.302S > Recept Ap.202S	69	72	3
	Floor system—Suite Emiss Ap.302 s > Recept Ap.202 s	64	69	5
	Floor system—Suite Emis Ap.302S > Recept Ap.202S	61	71	10
	Floor system—Suite Emis Ap.202S > Recept Ap.102S	67	69	2
I-10 (D9)	Floor system—Kitchen Emiss Ap.102 > Recept Ap.02	80	84	4

(continued)

Table 19 (continued)

Criterion (dB) (NBR 15575-3)	Environment	In-field	Numerical	Δ (dB)
Minimum: 66 to 80 Intermediate: 56 to 65 Superior: ≤55	Floor systems by separateing autonomous housing units into distinct floors	Result (dB) L'nT,w	Result (dB) L'nT,w	SC—MC
	Floor system—Kitchen Emiss Ap.103 > Recept Ap.03	83	84	1
	Floor system—Living room Emiss Ap.102 > Recept Ap.02	78	80	2
	Floor system—Living room Emiss Ap.103 > Recept Ap.03	77	80	3
	Floor system—Bedroom 1 Emiss Ap.101 > Recept Ap.01	79	84	5
	Floor system—Bedroom 1 Emiss Ap.102 > Recept Ap.02	80	84	4
	Floor system—Bedroom 1 Emiss Ap.103 > Recept Ap.03	82	84	2
	Floor system—Bedroom 1 Emiss Ap.104 > Recept Ap.04	82	84	2

(continued)

Table 19 (continued)

Criterion (dB) (NBR 15575-3)	Environment	In-field	Numerical	Δ (dB)
Minimum: 66 to 80 Intermediate: 56 to 65 Superior: ≤55	Floor systems by separateing autonomous housing units into distinct floors	Result (dB) L'nT,w	Result (dB) L'nT,w	SC—MC
I-11 (D9)	Floor system—Kitchen Emiss Ap.101 > Recept Ap.01	80	83	3
	Floor system—Kitchen Emiss Ap.102 > Recept Ap.02	76	83	7
	Floor system—Living room Emiss Ap.101 > Recept Ap.01	78	80	2
	Floor system—Living room Emiss Ap.102 > Recept Ap.02	77	80	3
	Floor system—Bedroom 1 Emiss Ap.101 > Recept Ap.01	81	83	2
	Floor system—Bedroom 1 Emiss Ap.102 > Recept Ap.02	80	83	3
	Floor system—Bedroom 2 Emiss Ap.101 > Recept Ap.01	81	82	1

(continued)

present better values than field measurement, since in the computational simulation the conditions for calculating the sound insulation are predictable and considered ideal, as opposed to field measurement conditions, the unpredictability of external factors may negatively interfere with the expected results.

Table 19 (continued)

Criterion (dB) (NBR 15575-3)	Environment	In-field	Numerical	Δ (dB)
Minimum: 66 to 80 Intermediate: 56 to 65 Superior: ≤55	Floor systems by separateing autonomous housing units into distinct floors	Result (dB) L'nT,w	Result (dB) L'nT,w	SC—MC
	Floor system—Bedroom 2 Emiss Ap.102 > Recept Ap.02	79	82	3
I-12 (D9)	Floor system—Kitchen Emis Ap.203 > Recept Ap.103 TB	80	86	6
	Floor system—Kitchen Emis Ap.304 > Recept Ap.204TA	80	86	6
	Floor system—Living room Emis Ap.203 > Recept Ap.103 TB	82	83	1
	Floor system—Living room Emis Ap.304 > Recept Ap.204TA	81	83	2
	Floor system—Bedroom 1 Emis Ap.203 > Recept Ap.103 TB	74	83	9
	Floor system—Bedroom 1 Emis Ap.304 > Recept Ap.204TA	75	84	9

(continued)

For SVH-RI, the lower the result, the better the bedroom to impact noise, because the L'_{nTw} index refers to the standard impact sound pressure level that is transmitted to the adjacent environment. Therefore, if SC presents values higher than MC, it means that the measurement obtained a better result than expected in the simulation, contrary to the initial hypothesis.

Table 19 (continued)

Criterion (dB) (NBR 15575-3)	Environment	In-field	Numerical	Δ (dB)
Minimum: 66 to 80 Intermediate: 56 to 65 Superior: ≤55	Floor systems by separateing autonomous housing units into distinct floors	Result (dB) L'nT,w	Result (dB) L'nT,w	SC—MC
	Floor system—Bedroom 2 Emis Ap.203 > Recept Ap.103 TB	73	84	11
	Floor system—Bedroom 2 Emis Ap.304 > Recept Ap.204TA	75	85	10
I-13 (D9)	Floor system—Bedroom Emiss Ap.201 > Recept Ap.101	82	82	0
	Floor system—Bedroom Emiss Ap.204 > Recept Ap.104	80	82	2
	Floor system—Suite Emiss Ap.201 > Recept Ap.101	81	83	2
	Floor system—Suite Emiss Ap.204 > Recept Ap.104	83	83	0
	Floor system—Living room Emiss Ap.201 > Recept Ap.101	80	77	−3

(continued)

4.2 Summary of the Results Presented

Table 20 presents, synthetically, the classification adopted in the present study for all the results of the comparative SC versus MC in relation to the variation of the difference Δ(dB) calculated SC-MC.

Table 19 (continued)

Criterion (dB) (NBR 15575-3)	Environment	In-field	Numerical	Δ (dB)
Minimum: 66 to 80 Intermediate: 56 to 65 Superior: ≤55	Floor systems by separateing autonomous housing units into distinct floors	Result (dB) L'nT,w	Result (dB) L'nT,w	SC—MC
	Floor system—Living room Emiss Ap.204 > Recept Ap.104	79	77	−2
I-14 (D9)	Floor system—Living room Emiss Ap.203 > Recept Ap.103	78	82	4
	Floor system—Kitchen Emiss Ap.203 > Recept Ap.103	80	86	6
	Floor system—Bedroom 1 Emiss Ap.203 > Recept Ap.103	77	84	7

Fig. 34 Scatter plot of the results (L'nT,w) measured in field and simulated—SVH-RI

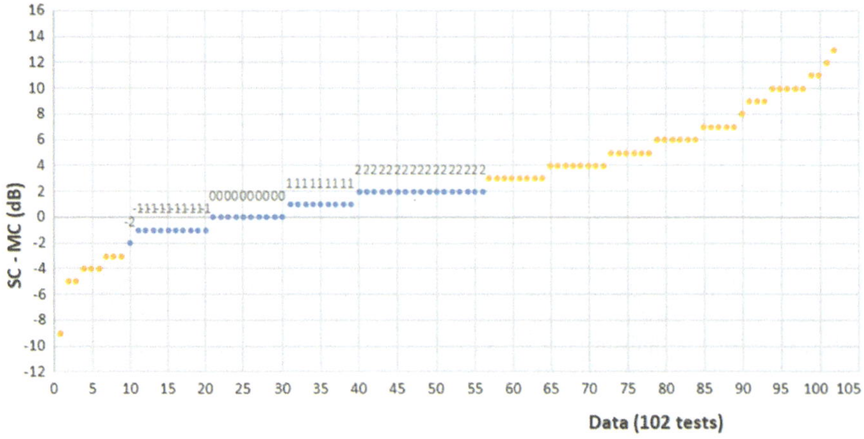

Fig. 35 Dispersion results of SC-MC (in dB)—SVH-RI geral

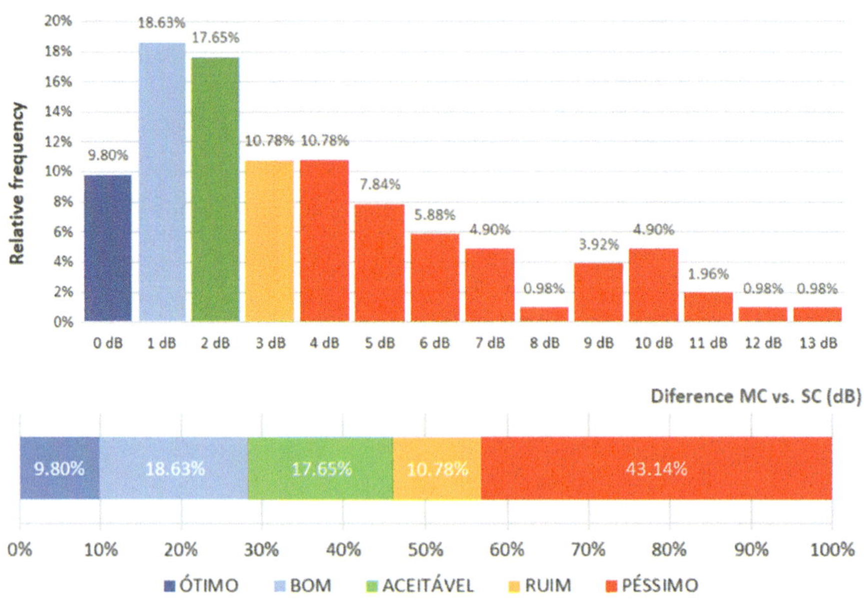

Fig. 36 Relative frequency histogram of the SC versus MC difference for general SVH-RI geral

Regarding the distancing of SC and MC, Table 21 presents a summary of the comparison of the SC versus MC indices for all systems studied, i.e., how much the SC results were higher, lower or equal to the results of MC.

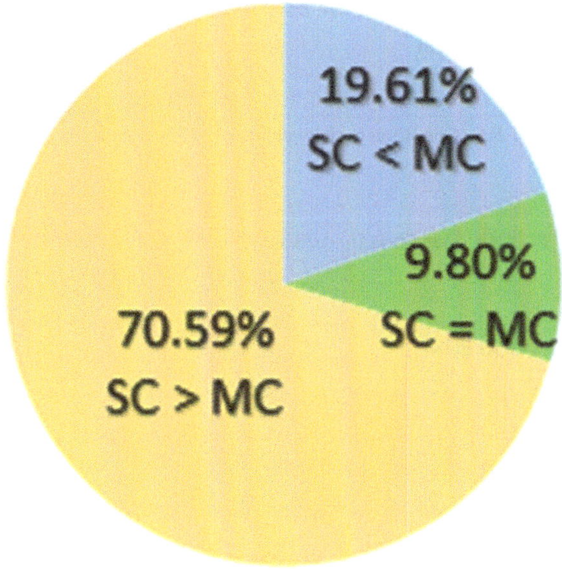

Fig. 37 Comparison of SC versus MC, for the SVH-RI results

Table 20 Classification summary of the systems studied

SVH-RA geral	**37.93%**	Optimal until Acceptable **0 dB-2 dB**
	62.07%	Bad to Poor: **3 dB and > 4 dB**
SVH-RA (Condition I—between floors with bedrooms)	43.08%	Optimal until Acceptable **0 dB-2 dB**
	56.92%	Bad to Poor: **3 dB and > 4 dB**
SVH-RA (Condition II—between floors without bedrooms)	31.37%	Optimal until Acceptable **0 dB-2 dB**
	68.63%	Bad to Poor: **3 dB and > 4 dB**
SVH-RI geral	**46.08%**	Optimal until Acceptable **0 dB-2 dB**
	53.92%	Bad to Poor: **3 dB and > 4 dB**

4.3 Specific Situations—Discussions of the Particularities

During the computational simulation stage, some specificities were observed in the evaluation of the expected results. The most important observation concerns the input data, which is essential for the reliability of the simulation step. Input data that differs from actual execution data will produce incompatible results.

In some case studies, the simulation was redone after observing a great distance between the SC and MC results, since the results of MC were already known. Now the data have been ratified, now they have been corrected. Thus, there were situations

Table 21 Comparative summary (Distance SC vs. MC)

SVH-RA geral	**48.28%** **1.72%** **50.00%**	SC > MC SC = MC SC < MC
SVH-RA (Condition I—between floors with bedrooms)	24.62%	SC > MC
	3.08%	SC = MC
	72.31%	SC < MC
SVH-RA (Condition II—between floors without bedrooms)	78.43%	SC > MC
	0%	SC = MC
	21.57%	SC < MC
SVH-RI geral	**70.59%**	SC > MC
	9.80%	SC = MC
	19.61%	SC < MC

in which the new simulation did not present any difference from the initial one, but there were others in which some particularities required remodeling that generated new bedroom values, which revealed some specificities.

4.3.1 SVH in the Cases of Solid Concrete Walls and Slabs—Calibration of the Data

An adjustment was made for the simulations of the floor systems—SVH of the concrete elements, calibrating the standardized damping factor from 0.003 to 0.009 and then remade the models in Insul and then applied the calculated results of sound reduction to the elements in the computational simulation of the building in order to evaluate the acoustic behavior of the environments that had the field measurements and enable, therefore a comparative analysis.

For the cases of VHS, with the adjustment of the damping, it was observed that for the air noise bedroom tests (SVH-RA), there was generally an increase of only 1 dB in the values resulting from the simulation.

For the simulation in relation to the bedroom to impact noise (SVH-RI), the adjustment in the damping favored a sound reduction in the order of 4 dB, although almost all the results in the simulation, for this condition, were higher than the field measurement, contrary to the initial hypothesis that the simulation should result in better values in relation to the values of the field measurement, which for air noise insulation would import smaller simulation values than those measured in the field.

Table 22 and Figs. 38 and 39 present the results of field measurement and computational simulation with the variation of damping used in the comparative analysis of SVH-RI data for case studies I-10 to I-14.

Using the standardized damping of 0.003 for the solid concrete elements, the sample analyzed presented a maximum value of 16 dB for the SC-MC difference, with an average of 7.8 dB suggesting for these cases incompatibility between simulation

Table 22 SVH-RI results for concrete slab study cases (particular data calibration cases—damping adjustment)

Environment		In-field	Numerical Damp. 0.003	Numerical Damp. 0.009	Δ (dB)	Δ (dB)
		Result (dB) $L_{nT,w}$	*Result (dB)* $L_{nT,w}$	*Result (dB)* $L_{nT,w}$	*SC1 -MC*	*SC2 -MC*
I-10	Floor System—Kitchen Emission Ap.102 > Recept Ap.02	80	89	84	9	4
	Floor System—Kitchen Emission Ap.103 > Recept Ap.03	83	89	84	6	1
	Floor System—Living room Emission Ap.102 > Recept Ap.02	78	85	80	7	2
	Floor System—Living room Emission Ap.103 > Recept Ap.03	77	85	80	8	3
	Floor System—Bedroom 1 Emission Ap.101 > Recept Ap.01	79	88	84	9	5
	Floor System—Bedroom 1 Emission Ap.102 > Recept Ap.02	80	88	84	8	4
	Sistema de Piso—Bedroom 1 Emission Ap.103 > Recept Ap.03	82	88	84	6	2
	Sistema de Piso—Bedroom 1 Emission Ap.104 > Recept Ap.04	82	88	84	6	2
I-11	Floor System—Kitchen Emission Ap.101 > Recept Ap.01	80	87	83	7	3

(continued)

Table 22 (continued)

Environment		In-field	Numerical Damp. 0.003	Numerical Damp. 0.009	Δ (dB)	Δ (dB)
		Result (dB) $L_{nT,w}$	*Result (dB)* $L_{nT,w}$	*Result (dB)* $L_{nT,w}$	*SC1 -MC*	*SC2 -MC*
	Floor System—Kitchen Emission Ap.102 > Recept Ap.02	76	87	83	11	7
	Floor System—Living room Emission Ap.101 > Recept Ap.01	78	84	80	6	2
	Floor System—Living room Emission Ap.102 > Recept Ap.02	77	84	80	7	3
	Floor System—Bedroom 1 Emission Ap.101 > Recept Ap.01	81	87	83	6	2
	Floor System—Bedroom 1 Emission Ap.102 > Recept Ap.02	80	87	83	7	3
	Floor System—Bedroom 2 Emission Ap.101 > Recept Ap.01	81	86	82	5	1
	Floor System—Bedroom 2 Emission Ap.102 > Recept Ap.02	79	86	82	7	3
I-12	Floor System—Kitchen Emiss Ap.203 > Recept Ap.103 TB	80	90	86	10	6

(continued)

and measurement in SVH-RI. With the adjustment in the damping of the element, the maximum value of this calculated difference fell to 11 dB and average in the order of 3.5 dB, suggesting that the calibration in the simulation will allow to achieve simulated results potentially closer to those measured in the field.

Table 22 (continued)

Environment		In-field	Numerical Damp. 0.003	Numerical Damp. 0.009	Δ (dB)	Δ (dB)
		Result (dB) $L_{nT,w}$	*Result (dB)* $L_{nT,w}$	*Result (dB)* $L_{nT,w}$	*SC1 -MC*	*SC2 -MC*
	Floor System—Kitchen Emiss Ap.304 > Recept Ap.204 TA	80	90	86	10	6
	Floor System—Living room Emiss Ap.203 > Recept Ap.103 TB	82	87	83	5	1
	Floor System—Living room Emiss Ap.304 > Recept Ap.204 TA	81	87	83	6	2
	Floor System—Bedroom 1 Emiss Ap.203 > Recept Ap.103 TB	74	88	83	14	9
	Floor System—Bedroom 1 Emiss Ap.304 > Recept Ap.204 TA	75	88	84	13	9
	Floor System—Bedroom 2 Emiss Ap.203 > Recept Ap.103 TB	73	89	84	16	11
	Floor System—Bedroom 2 Emiss Ap.304 > Recept Ap.204 TA	75	89	85	14	10
I-13	Floor System—Bedroom Emiss Ap.201 > Recept Ap.101	82	86	82	4	0

(continued)

Even adjusting this condition of element damping, the results of SC for SVH-RI in the solid concrete slabs analyzed were worse than THE, in the order of 87.88%, allowing us to infer that the choice of the use of computational simulation of acoustic behavior of SVH-RI in solid concrete slabs will result in worse insulation values as opposed to those measured in the field, thus allowing us to work for safety, because

Table 22 (continued)

Environment		In-field	Numerical Damp. 0.003	Numerical Damp. 0.009	Δ (dB)	Δ (dB)
		Result (dB) $L_{nT,w}$	*Result (dB)* $L_{nT,w}$	*Result (dB)* $L_{nT,w}$	*SC1 -MC*	*SC2 -MC*
	Floor System—Bedroom Emiss Ap.204 > Recept Ap.104	80	86	82	6	2
	Floor System—Suite Emiss Ap.201 > Recept Ap.101	81	87	83	6	2
	Floor System—Suite Emiss Ap.204 > Recept Ap.104	83	87	83	4	0
	Floor System—Living room Emiss Ap.201 > Recept Ap.101	80	81	77	1	−3
	Floor System—Living room Emiss Ap.204 > Recept Ap.104	79	82	77	3	−2
I-14	Floor System—Living room Emiss Ap.203 > Recept Ap.103	78	86	82	8	4
	Floor System—Kitchen Emiss Ap.203 > Recept Ap.103	80	90	86	10	6
	Floor System—Bedroom 1 Emiss Ap.203 > Recept Ap.103	77	88	84	11	7

in the simulation will be sought the minimum level of the performance standard, and in the field, potentially better values will be measured. Figure 39 shows the total comparison of the distance between the results of SC and MC for these specific cases of SVH-RI in concrete slabs.

The calibration of the input data in the modeling phase allowed to prove that the assertiveness (objectivity/clarity/safety) of these data is of fundamental relevance to the credibility of computational simulation as an acoustic prediction tool in the

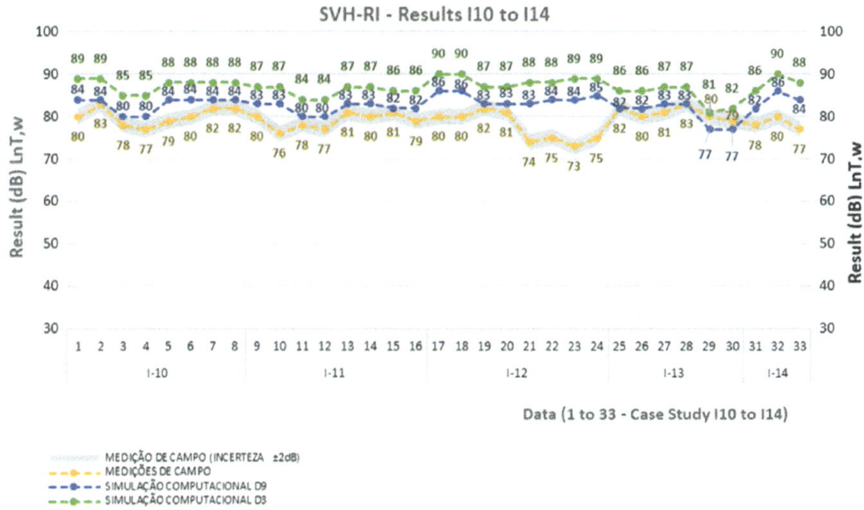

Fig. 38 Particular case—Adjustment in concrete damping—SVH-RI for I-10, I-11, I-12, I-13 and I-14

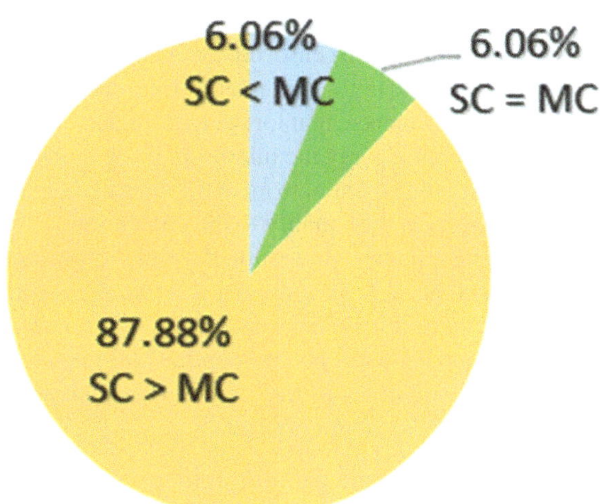

Fig. 39 Comparison of results SC versus MC—SVH-RI (Slabs of solid concrete)

projects. Otherwise, dubious data suggest illegitimate results, proving the expected reliability of this process.

5 Conclusions

From the analyses of the results of the case study of this experimental work, it can be inferred that the computational simulation of acoustic behavior presents potential for compatibility with field measurement of around 38% for the evaluation system of horizontal partition for air noise insulation (SVH-RA) and in the order of 46% for the horizontal partition evaluation systems for insulation at impact noise (SVH-RI).

This compatibility is linked to the percentage of values of the Difference SC versus MC that were within the range classified as acceptable of ±2 dB, referring to the margin of uncertainty of the field measurement provided by NBR ISO 16283-1 (2018) and ISO 12999-1 (2014), which induces an interpretation that the values within this range are potentially equal.

It was observed that, although a portion of the comparative analysis showed a distance above 3 dB between field measurement and computational simulation, by which the criterion of classification of values of this order was adopted as bad results, since they were outside the margin of uncertainty of ±2 dB, this portion from 3 dB allowed two interpretations:

- in cases where the computer simulation was lower than the SC < MC field measurement, it is admitted to suggest that when using computer simulation, it is working in favor of safety, with great possibilities of not only achieving the predicted, but of overcoming it, thus ensuring a higher quality than that required in the standard. It means that, if in the simulation the enterprise was predicted to meet a certain level of performance, that for the requirement of sound insulation the performance standard requires only the minimum level, in the field will be found a better insulation, with the potential to meet the intermediate or higher levels of said standard, thus giving a higher quality to the enterprise;
- in cases where the computational simulation was greater than the SC-MC in-field measurement, it is then proven the initial hypothesis that the execution uncertainties, the construction failures and the potential field interferences decrease the acoustic bedroom potential of the building elements, which were predicted in controlled situations of the computational simulation, which are the same ideal situations of the laboratory tests.

Regarding horizontal partition systems, the analysis of air noise bedroom (SVH-RA) of the total data showed no tendency for the simulation to be higher or lower than the measurement, since the percentage of SC < MC and SC > MC were, respectively, 50.0% and 48.3%, considered practically equal. However, analyzing separately the types of environments, for the case in which the between floors systems was between dormitories, the analysis of the data pointed to a trend of the values of the computational simulation being lower than the data measured in the field, making in practice better results than the minimum level required in the performance standard, which will be as predicted in the simulation.

Regarding the analysis of impact noise, the comparison of the data revealed that 70% of the computational simulation resulted in values worse than the field measurement, that is, SC > MC, which for this condition the simulation should, as an initial hypothesis, present values lower than the measurement, since the verification is of the noise that passes from one environment to another. This said, it is concluded that using the simulation data in the projects will enable better quality after the execution.

In the particular cases of solid concrete elements, both walls and slabs, the study allowed us to conclude that the calibration of input data in the systems are crucial for the reliability of the simulation. The variation of data such as specific density in the case of wall analysis, and damping in the case of solid slabs, allowed to verify that:

- for the simple concrete walls, the calibration allowed in the modeling to reach the indicative values of 45 dB found in previous research for the concrete walls of 10 cm thick, and thus model the elements of the projects based on the specific input data of each project, however, compatible with the models tested in the laboratory by other researchers, to then evaluate the overall acoustic behavior of the study cases of this work;
- for the floors, the calibration of the damping rate allowed an average increment of 1 dB in airborne noise insulation and 4 dB in the bedroom to impact noise. However, this increase of 1 dB for the SVH-RA system caused the portion of data within the acceptable range (considered up to 2 dB) to drop from 41% to 38%, while for SVH-RI, the adjustment in damping allowed an approximation of the results within the acceptable range, which was 32% in damping from 0.003 and increased to 46% when considered 0.009. However, even after the increment, the simulation presented worse results than those measured in the field, with greater absolute difference in cases of SVH-RI.

Thus, it is concluded as positive the use of computer simulation as a tool to predict the acoustic behavior of real estate, as it allows to predict, before the execution of the work, how the enterprise will behave in terms of soundproofing, adjusting, in the design phase, the type and thickness of elements, the material and characteristics of these, whose new solutions favor better damping and enhance the mass/spring/mass system for high frequency sounds or greater mass to attenuate the low frequency sounds, thus aiming at meeting the expected performance levels.

And, considering the possibility of the measurement presenting values better than the simulated values, that is, contrary to the expected simulation present better results due to the ideal conditions to the detriment of field interferences that would tend to worsen the expected results, it is understood that this circumstance will enhance the range of better indexes than the minimum required in the performance standard, allowing the user to receive and enjoy a building with a higher quality than expected.

5.1 Suggestions for Future Work

Following this study, some analyses may be made regarding the specificities found in this sample.

It is suggested to perform the comparative analysis with more similar types of construction systems and with a greater amount of sample data, in order to allow statistical treatment.

It is also suggested that the laboratory measurement of the elements of the construction systems adopted be made, according to the method of constant precision of NBR 15575 (2013), with the purpose of better representativeness of the input data used in the modeling, which will allow, therefore, greater assertiveness in the computational simulation of the acoustic behavior of the studied building compared to the veracity of its executiono.

It is suggested to evaluate the types of systems separately, i.e. only ceramic masonry systems, only concrete block masonry systems, only solid concrete wall systems, only ribbed slabs, only solid slabs, only precast slabs, etc.

In view of horizontal partition systems, it is suggested to evaluate how much floor covering can determine improvement in the level of sound insulation.

As the properties had already been delivered to the owners, it was not possible to redo some measurements that suggested measurement error, in this sense, it is suggested for the next work that this comparison of simulation in properties with field measurement, allow the simulation to be done in parallel to the completion of the work to allow confirmation of dubious values before the occupation of the properties.

Also in relation to this aspect of occupation, the study can be expanded by evaluating the difference in field measurement results in empty properties and in occupied properties, to analyze how much occupation impacts the expected performance of sound insulation.

References

Agopyan, V., & John, V. M. (2011). *The challenge of sustainability in civil construction* (Vol. 5). Blucher, Brazil: São Paulo.

Almeida, M., Silva, S., & Ferreira, T. (2006). Building physics. *Environmental and Building Acoustics*, Universidade do Minho, Portugal.

Arndt, L. T., Philips, J. W., & Barbosa, W. A. (2010). Geographic information system for urban noise mapping. Article. In: *III Brazilian Symposium on Geodesic Sciences and Geoinformation Technologies*. Recife: SIMGEO, Brazil.

Brazilian Association for Acoustic Quality (ProAcústica). (2017a). ProAcoustic Manual on Performance Standard, São Paulo, Nrazil.

Brazilian Association for Acoustic Quality (ProAcústica). (2017b). ProAcoustic Manual for noise class of housing buildings, São Paulo, Nrazil.

Berrio, W. B., & Castro, S. S. (2012). *Sound impact assessment at the Pontifical Catholic University of Peru*. Ph.D. Thesis, Lima: Pontificia Universidad Católica del Peru, Peru.

Bezerra, A. F. D., Brito, A. C., Aquilino, M. M., Akutsu, M., & Vittorino, F. V. (2018). Electrical outlet boxes and wall insulation failures. *Proceedings of the National Meeting of Built Environment Technology*, 17, 2018, Foz do Iguaçu, Porto Alegre: ANTAC, Brazil.

BORGES, C. A. M. The concept of building performance and its importance for the construction sector in Brazil. MSc Thesis, Escola Politécnica da Universidade de São Paulo, São Paulo, Brazil, 2008.

Borges, C. A. M., & Sabbatini, F. H. (2008). The concept of building performance and its importance for the construction sector in Brazil. Technical Bulletin of the Polytechnic School of USP, Department of Civil Engineering, BT/PCC/515. Universidade de São Paulo, São Paulo, Brazil.

Boselli, T., & Dunowicz, R. (2009). La calidad y la conservación de la vivienda social: Bases para el aseguramiento de la sustentabilidad del hábitat. *In.* V National Meeting and III Latin American Meeting on Sustainable Buildings and Communities. Recife, Brazil (in Spanish).

Boufleur, V. (2013). *Acoustic performance of housing buildings: challenges for implementation of the performance standard*. M.Sc. Thesis, Universidade Federal do Rio Grande do Sul, Porto Alegre, Brazil.

Brazilian Association of Technical Standards (ABNT). (2013). NBR 15575: Housing Buildings—Performance, Rio de Janeiro, Brazil.

Brazilian Association of Architecture Offices (ASBEA). (2015). Guide for Architects in applying the Performance Standard NBR 15.575. São Paulo, Brazil.

Brazilian Service of Support to Micro and Small Companies (SEBRAE). (2016). Prospective Scenarios: The construction sector in Brazil, from 2016 to 2018. SEBRAE, Brazil.

Brazilian Association of Technical Standards (ABNT). (2017). NBR 10152: Acoustics — Sound pressure levels of indoor environments, Rio de Janeiro, Brazil.

Brazilian Chamber of Construction Industry (CBIC). (2013). Performance of housing buildings: guiding guide to meet the standard ABNT NBR 15575/2013. Fortaleza: Gadioli Cipolla Comunicação, Brazil.

Brazilian Association of Technical Standards (ABNT). (2019). NBR 10151: Acoustics—Measurement and evaluation of sound pressure levels in inhabited environments—Application for general use, Rio de Janeiro, Brazil.

Carraro, F., & Jucá, T. (2014). Performance Standard smells like opportunity. Revista Construir Mais—SINDUSCON. Goiás, Brazil.

Carvalho, L. N. (2015). Impact noise on slabs: Comparative analysis between performance of computational models and in-field tests. MSc Thesis, USP, São Paulo, Brazil.

Catai, R. E., Penteado, A. P., & Dalbello, P. F. (2006). Materials, techniques and processes for sound insulation. In: *17 Brazilian Congress of Engineering and Materials Science—CBECiMat*, Foz do Iguaçu, PR, Brazil.

Cotta, A. C., & Andery, P. R. P. (2018). Changes in the design process of construction and developer companies due to NBR 15575—Performance Standard. *Ambiente Construído, 18*(1), 1–15.

Crusius, A. D. (2015). Methodology for implementation of abnt nbr 15575/2013—Performance of housing buildings—in construction companies and architectural projects focused on the real estate market. *Revista Especialize, 1*(10), 1–18.

Ferreira, J. A. C. (2004). *Acoustic evaluation of residences: measured and simulated results*. M.Sc. Thesis, Universidade Federal do Paraná, Curitiba, Brazil.

Ferreira, J. A. C., & Zannin, P. H. T. (2007). Determination of sound insulation coefficients: in situ measurements and numerical simulation. *Ambiente Construído, 7*(2), 15–29.

Freitas, R. M. (2005). Between myths and limits: the possibilities of constructive density in the face of quality of life in the urban environment, PhD Thesis, Universidade Federal do Rio Grande do Sul. Porto Alegre, Brazil.

Gibson, E. J. (1982). Working with the Performance Approach in Building, CIB Report Publication n. 64, Rotterdam.

Giunta, M. B. (2013). *Analysis of acoustic forecast modeling and sound mapping for the city of São Carlos*, M.Sc. Thesis, Universidade Federal de São Carlos. São Carlos, Brazil.

International Organization for Standardization. (2004). *10052: Acoustics—Field measurements of airborne and impact sound insulation and of service equipment sound—Survey method*. Switzerland: Genève.

International Organization for Standardization. (2008). *ISO 3382-2: Acoustics — Measurement of room acoustic parameters—Part 2: Reverberation time in ordinary rooms*. Switzerland: Genève.

International Organization for Standardization. (2010). *ISO 10140-2: Acoustics—Laboratory measurement of sound insulation of building elements—Part 2: Measurement of airborne sound insulation*. Switzerland: Genève.

International Organization for Standardization. (2013). *ISO 717-1: acoustics: rating of sound insulation in buildings and of building elements: Part 1: Airborne sound insulation*. Switzerland: Genève.

International Organization for Standardization. (2014). *ISO 12999-1: Acoustics — Determination and application of measurement uncertainties in building acoustics — Part 1: Sound insulation*. Switzerland: Genève.

International Organization for Standardization. (2017a). *ISO 12354-1: Building acoustics—Estimation of acoustic performance of buildings from the performance of elements—Part 1: Airborne sound insulation between rooms*. Switzerland: Genève.

International Organization for Standardization. (2017b). *ISO 12354-2: Building acoustics — Estimation of acoustic performance of buildings from the performance of elements — Part 2: Impact sound insulation between rooms*. Switzerland: Genève.

International Organization for Standardization. (2017c). *ISO 12354-3: Building acoustics — Estimation of acoustic performance of buildings from the performance of elements — Part 3: Airborne sound insulation against outdoor sound*. Switzerland: Genève.

International Organization for Standardization. (2017d). *ISO 16283-1: Acoustics — Field measurement of sound insulation in buildings and of building elements — Part 1: Airborne sound insulation — Amendment 1*. Switzerland: Genève.

International Organization for Standardization. (2018). *ISO 16283-2: Acoustics — Field measurement of sound insulation in buildings and of building elements — Part 2: Impact sound insulation*. Switzerland: Genève.

Jensen, H. A. R., Rasmussen, B., & Ekholm, O. (2019). Neighbour noise annoyance is associated with various mental and physical health symptoms: results from a nationwide study among individuals living in multi-storey housing. BMC Public Health, v. 19.

Kern, A. P., Silva, A., & Kazmierczak, C. S. (2014). The process of implementation of performance standards in construction: A comparison between Spain (CTE) and Brazil (NBR 15575/2013). *Gestão e Tecnologia de Projetos, 9*(1), 89–102.

Mahmood, S. F. (2017*). From stilts to housing estates: a study on thermal comfort in the set Via Mangue II, Recife—PE*. M.S.c Thesis, Universidade Federal de Pernambuco, Recife, Brazil.

Michalski, R. L. X. N. (2011). *Methodologies for measuring sound insulation in the field and for the expression of measurement uncertainty in the evaluation of the acoustic performance of buildings*. Ph.D. Thesis, Universidade Federal do Rio de Janeiro, Rio de Janeiro, Brazil.

Morais, G.; Lira, V.; Pereira, A.; Nascimento, E.; Lordsleem Jr., A.; Melhado, S. Acoustic performance and architectural design companies in light of the performance standard. In: National Meeting of Built Environment Technology, 17, Foz do Iguaçu, Porto Alegre: ANTAC, Brazil, 2018.

Moreira, F. S., Lima, R. F., & Cândido, L. F. (2018). The challenges for the implementation of acoustic performance requirements according to ABNT NBR 15.575/2013: A case study with construction companies in northeastern Brazil. In: National Meeting of Built Environment Technology, 17, Foz do Iguaçu, Porto Alegre: ANTAC, Brazil.

Neto, M. F. F. (2009). *Acoustic comfort level: A proposal for residential buildings*. Ph.D. Thesis, Universidade Estadual de Campinas, Campinas, Brazil, 2009.

Neto, M. De F. F., Bertoli, S. R., & Barry, P. J. (2010). Difference between acoustic performance tests in laboratory and field in masonry walls. In *XXIII Meeting of the Brazilian Society of Acoustics, Salvador*, Brazil.

Official Journal of the European Communities (OJEC). Directive 2002/49/EC of European planning and the Council. June 25, 2002. Available at: http://eur-lex.europa.eu/LexUriServ/LexUriServ.do?uri=OJ:L:2002:189:0012:0025:PT:PDF. Access 18 Mar. 2019.

Paixão, D. X. (1997). *Analysis of acoustic conditions in the classroom*. M.S.c Thesis, Universidade Federal de Santa Maria, Rio Grande do Sul, Brazil.

Patricio, J. (2005). Building acoustics: insulation index to percussion sounds used in European space. *Revista de Acústica, 36*(1–2), 1–15.

Patricio, J. (2010). *Acoustics in buildings* (6th ed.). Lisboa: Verlag Dashofer, Germany.

Piffer, V. S. (2013). *Service life and performance of constructions in the NBR 15575 (2013)*. São Paulo, Brazil: USP.

Rasmussen, B. (2004). Sound Insulation between dwellings—Classication schemes and buildig regulations in Europe. In *The international Congress and Expositionon Noise Control Engineering*, 33, Praga.

Resende, J. M. S., Morais Filho, J. C. G., & Nascimento, N. L. F. (2014). *Acoustic performance according to performance standard NBR 15575*. M.Sc. Thesis, Universidade Federal de Goiás, Goiânia, Brazil.

Ribas, R. A. J. (2013). *Method for evaluating the thermal and acoustic performance of buildings applied in industrialized closing panels*. Ph.D. Thesis, Universidade Federal de Ouro Preto, Ouro Preto, Brazil.

Santos Filho, V. M. D., Sposto, R. M., & Caldas, L. R. (2017). Analysis of the acoustic performance of ventilated porcelain facades in the light of the performance standard: A case study in a housing building in Brasilia. DF. *REEC—Revista Eletrônica de Engenharia Civil, 13*(2), 116–130.

Silva Junior, O. J. (2012). *Evaluation of sound intensity in buildings*. M.SC. Thesis, Universidade Federal de Pernambuco, Recife, Brazil.

Silva, A. T., Kern, A. P., Piccoli, R., & González, M. A. S. (2014). New requirements arising from building environmental certification programs and building performance standards. *Arquiteturarevista, 10*(2), 105–114.

Souza, L. C. De, Almeida, M. G. De, & Bragança, L. (2013). *Bê a bá* of architectural acoustics: listening to architecture. São Carlos: Edufscar, Brazil.

Takahashi, V. F. M. (2016). *Acoustic performance of buildings: computational tool for evaluation*. Ph.D. Thesisi, Universidade Estadual de Campinas, Campinas, Brazil.

Takahashi, V. F. M., & Bertoli, S. R. (2012). Relationship between acoustic attributes and architectural characteristics of concert halls applying shape synthesis method. *Ambiente Construído, 12*(1), 193–205.

Union of Architecture and Engineering (SINAENCO). (2015). The impacts of the Performance Standard in the Architecture and Consulting Engineering Sector. SINAENCO, São Paulo, Brazil.